# Basic Electrical and Instrumentation Engineering

**Scrivener Publishing**
100 Cummings Center, Suite 541J
Beverly, MA 01915-6106

*Publishers at Scrivener*
Martin Scrivener (martin@scrivenerpublishing.com)
Phillip Carmical (pcarmical@scrivenerpublishing.com)

# Basic Electrical and Instrumentation Engineering

P. Sivaraman, C. Sharmeela,
A. Thaiyal Nayagi,
and R. Mahendran

Scrivener Publishing

WILEY

This edition first published 2021 by John Wiley & Sons, Inc., 111 River Street, Hoboken, NJ 07030, USA and Scrivener Publishing LLC, 100 Cummings Center, Suite 541J, Beverly, MA 01915, USA

For more information about Scrivener publications please visit www.scrivenerpublishing.com.



**Wiley Global Headquarters**
111 River Street, Hoboken, NJ 07030, USA

For details of our global editorial offices, customer services, and more information about Wiley products visit us at www.wiley.com.



***Library of Congress Cataloging-in-Publication Data***

ISBN 9781119764465

Cover image: Courtesy of the Authors
Cover design by Kris Hackerott

Set in size of 11pt and Minion Pro by Manila Typesetting Company, Makati, Philippines

Printed in the USA

# Dedications

***Mr. P. Sivaraman*** *dedicating to his father Mr. A. Palanisamy (deceased), mother Mrs. P. Valarmathi, sister Mrs. P. Shanmuga Priya, Spouse Mrs. A. Gowri and Daughter baby S. Aathira.*

***Dr. C. Sharmeela*** *dedicating to her parents Mr. N.S.Chenniappan and Mrs. C. Kasturi, my brother Chandrasekar, Vanitha, Shakthi Chandrasekar, Subathra and my beloved friends for encouraging me and extending their full support in writing the book.*

***Ms. A. Thaiyal Nayagi*** *dedicating to her mother A.Mariyayee, sister A. sujatha and my brother's A.Azhagirisamy, A.saravanamani for their constant encouragement and support.*

***Mr. R. Mahendran*** *dedicating to his father, mother and sister.*

# Dedication

# Contents

# Foreword

The backbone of electrical engineering is the power system, and it is one of the first things that a student in EE needs to learn. There are quite a large number of text books for courses on basics of electrical and instrumentation; this one presents a very simple, concise and clean approach that should make the subject very easily accessible to students. It will be particularly useful for revision after class lectures, and self study.

The first chapter introduces the student to the basic concepts of AC and DC power, voltage, current and the main constituents of a power system. In the subsequent chapters, students are introduced to DC and AC machines, as well as the basic tenets of measurement and instrumentation in the context of power systems. Chapter 2 describes the construction and operation of single phase and three phase transformers, while Chapter 3 does the same for DC machines. Chapter 4 covers AC machines, both induction machines and synchronous machines. Chapter 5 provides a brief exposition on simple measuring instruments and their operation. This should serve as a useful first introduction to power systems, before referencing advanced literature.

The authors have nicely blended their academic foundation with some industrial insight to make this book relevant and direct. I commend them on their work.

**Nandini Gupta**
Professor
Department of Electrical Engineering
Indian Institute of Technology, Kanpur

# Acknowledgements

First and foremost thanks to the Almighty for his everlasting love throughout this endeavor.

**Mr. P. Sivaraman** expresses his sincere thanks to Mr. Balaji Sriram, Research Scholar, IIT Kanpur, D. Sathiya Moorty, Research Scholar, IIT Ropar, Mr. Upendran, Research Scholar, IIT Madras; Mr. Priyaranjan Satpathy, Research Scholar, ITER, SOA Deemed to be University, Bhubaneswar, Mr. S. Rajkumar, Executive, JLL, Bengaluru, Mr. K. Sasikumar, Electrical Engineer, Mott MacDonald, Noida, Mr. Muthukumaran, Director, TECH Engineering Services, Chennai, Mr. K. Balaji, Electrical Engineer, Sree Nandees Technologies, Chennai, Mr. Ravichandran, Andrew Yule and Dr. S. Logesh Kumar, Dean Electronics, Coimbatore for providing their technical support, figures, expert review and finalizing the contents.

**Dr. C. Sharmeela** expresses her sincere gratitude to her mentor Dr. D.P.Kothari, Former Director (i/c), IIT Delhi, her research supervisor Dr. M.R.Mohan, Former Professor, Anna University, Chennai and Dr.S. Chandramohan, Professor & Head, DEEE, Anna University, Chennai for their continuous support and encouragement in completing this book.

**Ms. A. Thaiyal Nayagi** expresses her sincere thanks to the management of Rane polytechnic college, Dean, Principal and Head of the department of Mechanical and all staff members for their kind support and encouragement.

**Mr. R. Mahendran** expresses his sincere thanks to his friends and family members for their kind support and encouragement.

# Acknowledgements

1

# Introduction to Electric Power Systems

The main parameters in electrical systems are voltage and current. The product of voltage and current gives a third parameter called power, and power consumed over some time duration is called energy [1]. The electric power system consists of power generation, transmission and distribution system [2]. Power is generated from two main sources, namely conventional energy sources and non-conventional energy sources. The conventional energy sources are non-renewable which gets depleted over a period of time, while non-conventional energy sources are non-depleting sources [3, 10, 30]. Most large-scale power plants are located in areas where the raw materials are available locally and generated power is transmitted over long distances for distribution. An electrical conducting medium is required in order to transfer the power from generating station to load center. This conducting medium is called as transmission system. Transformers and transmission lines are the main components of the transmission system; they are used to transfer the power from generating station to consumers (customers) at various operating voltage levels. Generation voltage of the conventional power plant typically ranges from 6.6 kV to 22 kV, and transmission voltage typically ranges from 110 kV to 765 kV. High voltage is stepped down to various voltage levels for different consumers depending upon the requirements and installed capacity.

## 1.1 Introduction

In general, electrical power used for commercial and residential purpose is mostly Alternating Current (AC). An AC voltage or current has the magnitude and direction which changes periodically with respect to time, unlike the Direct Current (DC) supply, which has constant magnitude with respect to time. Figure 1.1 shows the waveform of AC (voltage and current) and Figure 1.2 shows the waveform of DC (voltage and current).

P. Sivaraman, C. Sharmeela, A. Thaiyal Nayagi, and R. Mahendran. Basic Electrical and Instrumentation Engineering, (1–78) © 2021 Scrivener Publishing LLC

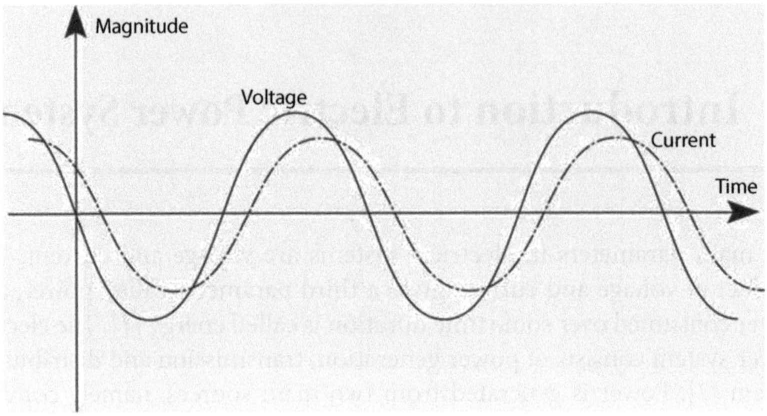


**Figure 1.1** Waveform of AC.

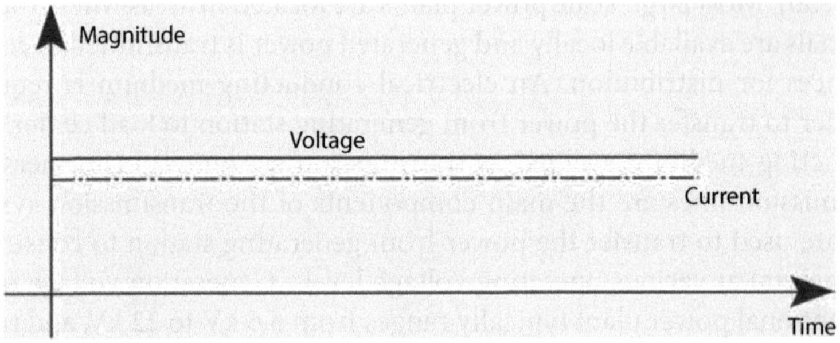


**Figure 1.2** Waveform of DC.

**Types of AC supply system:**

The AC supply system is classified into two types based on the number of phases:

- Single-phase power supply
- Three-phase power supply

**A. The single phase supply**

The single-phase power supply is used to power all the single-phase loads in the systems. Generally, single-phase power supply is derived from a three-phase, four-wire circuit. The single-phase voltage level varies from country to country. The general single-phase supply voltage is 220 V, 230 V, 240 V in low voltage systems (in Asian countries). Most of the single-phase supply systems are 2W systems as shown in Figure 1.11.

**B. Three-Phase Power Supply**

The three-phase power supply is used to power certain loads which need the poly-phase supply for their operation. Here phase means branch circuits or winding and poly means many. Such loads in any applications need a power supply which has a poly-phase supply system. For example, three-phase power supply is also called poly-phase power supply. In order to develop poly-phase supply, the armature winding of an alternator is divided into the number of phases as required. In each winding section, voltage gets induced with 120° displacement. These windings are arranged in such a way that the magnitude and frequency is the same for all the phases with definite phase difference with respect to the other phases. That means, in a three-phase power supply system, there are three voltages with equal magnitude and frequency having a phase difference of 360°/3 = 120° between them.

Advantages of three-phase supply systems are

- Single-phase power supply are obtained from three-phase power supply and in reverse three-phase power supply is not obtained from single-phase power supply
- Three-phase induction motors are self-starting motors where single-phase induction motors are not self-starting motors
- For transmission and distribution, a three-phase system needs smaller size conductor material as compared with single-phase system for same volt amperes

## 1.1.1 Electrical Parameters

The main parameters in an electrical system are voltage and current. The product of voltage and current gives a third parameter called power, and power consumed over some duration is called energy.

AC systems: Voltage, Current, Frequency and Phase angle

DC systems: Voltage and Current

### *1.1.1.1 Voltage*

Potential difference between any two points in an electrical circuit is called voltage. The SI unit of voltage is Volts (V) [1]. Higher values of voltage are mentioned as kV. The other name of voltage is Electro Motive Force (EMF). The representation of voltage is two types: peak to peak voltage (instantaneous voltage) and RMS voltage.

Alessandro Volta ($18^{th}$ Feb 1745 – $5^{th}$ March 1827): An Italian scientist who invented the first battery cell. In order to honour him, SI unit of electric potential is named Volt.

Alessandro Volta. Courtesy: Google image.

**RMS voltage:**

The peak to peak voltage of phase to phase is shown in Figure 1.3. Average voltage of positive half cycle to negative half cycle is zero. As the absolute voltage is not zero, average voltage cannot be used as a measuring scale for AC. In order to perform the analysis and calculation, a new term called RMS is considered for measurement. Theoretically, RMS voltage in AC is equivalent to the amount of heat produced if the DC of some magnitude produces the same heat on the same resistance.

Generally, most of voltage referred in specification is RMS voltage unless specified.

**Example 1.1:** A 40 W incandescent lamp is connected across 1ϕ, 230 V, 50 Hz AC supply.

Here the voltage 230 V is RMS voltage.

Note: When using the multi meter for voltage measurements, first check the meter is RMS rated or peak rated in order to avoid confusion. For an example, if RMS rated meter read the voltage as 230 V, the peak rated meter will read the same voltage as 325.2 V.

**RMS voltage from peak voltage:**

RMS voltage can be calculated if peak voltage is known. The expression for RMS voltage is given in eqn. 1.1.

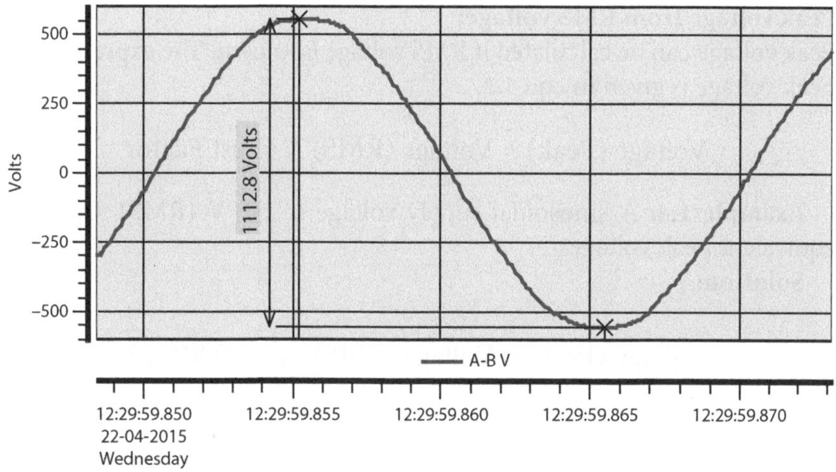


**Figure 1.3** Peak to peak voltage of R phase to Y phase. Note: This figure is captured using Dranetz Power Quality analyser.

$$\text{Voltage (RMS)} = \frac{\text{Voltage (Peak)}}{\text{Crest Factor}} \tag{1.1}$$

For sinusoidal wave shape, the value of crest factor is $\sqrt{2}$.

**Example 1.2:** A sinusoidal supply voltage is 340 V peak. Calculate equivalent RMS voltage.

**Solution:**

$$\text{Voltage (RMS)} = \frac{\text{Voltage (Peak)}}{\sqrt{2}} = \frac{340}{\sqrt{2}}$$

Voltage (RMS) = 240.5 V is ~241 V.

**Example 1.3:** A sinusoidal supply voltage is 565 V peak. Calculate equivalent RMS voltage.

**Solution:**

$$\text{Voltage (RMS)} = \frac{\text{Voltage (Peak)}}{\sqrt{2}} = \frac{565}{\sqrt{2}}$$

Voltage (RMS) = 399.6 V is ~400 V.

**Peak voltage from RMS voltage:**
Peak voltage can be calculated if RMS voltage is known. The expression for peak voltage is given in eqn 1.2.

$$\text{Voltage (Peak)} = \text{Voltage (RMS) X Crest Factor} \tag{1.2}$$

**Example 1.4:** A sinusoidal supply voltage is 240 V (RMS). Calculate equivalent peak voltage.

**Solution:**

$$\text{Voltage (Peak)} = \text{Voltage (RMS)} * \sqrt{2} = 240 * \sqrt{2}$$

$$\text{Voltage (Peak)} = 339.3 \text{ V is} \sim 339 \text{ V}$$

**Example 1.5:** A sinusoidal supply voltage is 415 V RMS. Calculate equivalent peak voltage.

**Answer:**

$$\text{Voltage (Peak)} = \text{Voltage (RMS)} * \sqrt{2} = 415 * \sqrt{2}$$

$$\text{Voltage (Peak)} = 586.8 \text{ V is} \sim 587 \text{ V.}$$

**Peak to peak voltage:**
Voltage measured between the maximum value of the positive half cycle and the minimum value of the negative half cycle is known as peak to peak voltage. This voltage is generally measured in individual voltage cycle, unlike the peak voltage which is the product of RMS voltage and crest factor measured in voltage trend. Peak to peak voltage of R to Y phase is shown in Figure 1.3.

The maximum voltage of positive half cycle of above waveform is 556.4 V and minimum voltage of negative half cycle of the waveform is 556.4 V. Voltage between positive half cycle to negative half cycle is +556.4 to -556.4 V is called peak to peak voltage. It is 1112.8 V in the above Figure 1.3. Figure 1.4 shows the peak voltage trend, which is different from peak to peak voltage.

From Figure 1.4, the peak voltage is between 555 V to 560 V for the time duration 70 minutes between 12:40 hours to 13:50 hours.

**Voltage in three-phase circuit:**
Voltage in a three-phase circuit is determined based on the system configuration. The determination of voltage in a circuit is based on whether the circuit is in star or delta configuration.

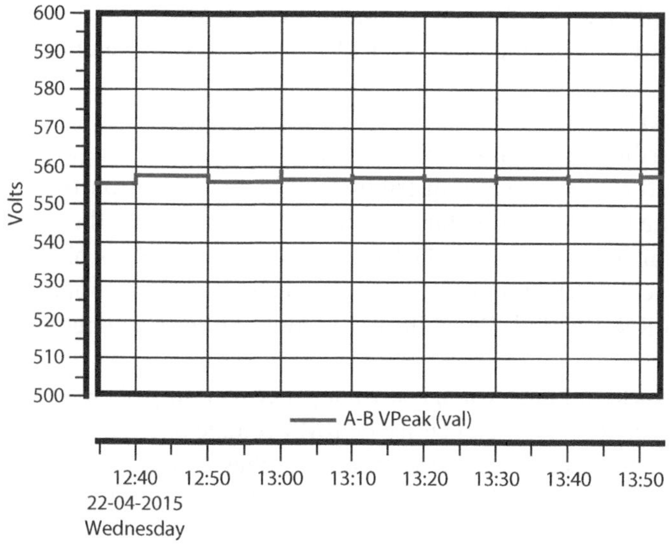


**Figure 1.4** Peak voltage trend. Note: This figure is captured using Dranetz Power Quality analyser.

1. Phase to phase voltage – For delta circuits
2. Phase to neutral voltage – For star circuits

The relationship between the above two voltage determinations is expressed in the equation (1.3) and (1.4).

In delta circuit, the phase voltage is determined by

$$\text{Voltage}_{\text{Ph}} = \frac{\text{Voltage}_{\text{Line}}}{\sqrt{3}} \tag{1.3}$$

In star circuit, the line voltage is determined by

$$\text{Voltage}_{\text{Line}} = \sqrt{3} \text{ x } \text{Voltage}_{\text{Ph}} \tag{1.4}$$

Where

$\text{Voltage}_{\text{Line}}$ is line voltage

$\text{Voltage}_{\text{Ph}}$ is phase voltage

In delta circuit, line current in delta circuit is greater than line current in star circuit. Similarly, the line voltage in star circuit is greater than line voltage in delta circuit.

**Example 1.6:** A distribution transformer of 2 MVA power rating Dyn11 configuration is having ratio of 11 kV/433 V. How do we understand this voltage?

**Answer:**

A distribution transformer is generally used to cater single-phase loads connected on three-phase distribution. This is the reason the secondary side of the transformer is star with neutral in the circuit. The primary side of the transformer is delta. 11 kV at primary side means phase to phase voltage across RY, YB and BR. 433 V secondary corresponds to 250 V between RN, YN and BN while 433 V could be recorded between RY, YB and BR.

The phase to phase voltage is voltage between any of two phases (R to Y, Y to B or R to B). It can either be RMS voltage or peak voltage and it is applicable to three-phase 3wire and three-phase 4wire circuits. The three-phase 4wire, 415 V, 50 Hz AC circuit is shown in Figure 1.5.

**Example 1.7:** What is the voltage between R to Y phase in Figure 1.5?
**Answer:** 415 V.

The three-phase 3W, 11 kV, 50 Hz AC circuit is shown in Figure 1.6.

**Example 1.8:** What is the voltage between Y to B phase in Figure 1.6?
**Answer:** 11 kV.

In India, the transmission voltages are 11 kV, 22 kV, 33 kV, 66 kV, 110 kV, 132 kV, 220 kV, 230 kV, 400 kV and 765 kV and the distribution voltage is

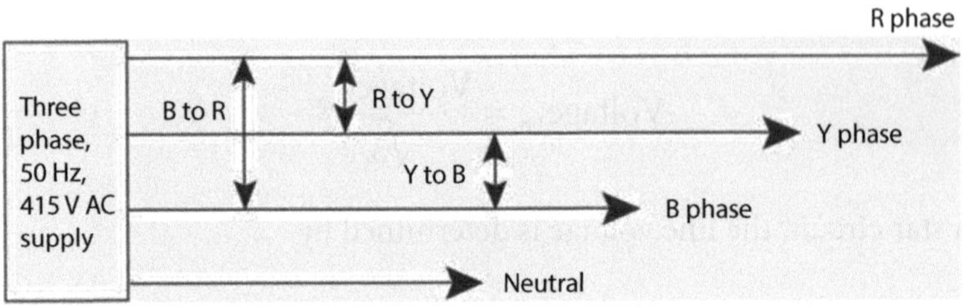


**Figure 1.5** Three-phase, Four-wire circuit configuration.

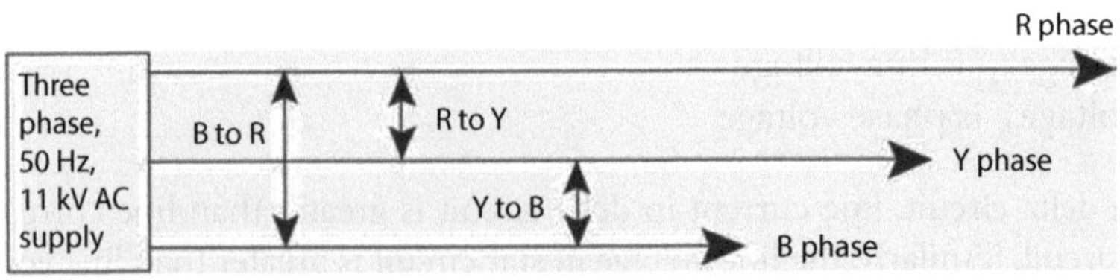


**Figure 1.6** Three-phase, Three-wire circuit configuration.

415 V. All the above-mentioned voltages are phase to phase voltage only. Similar voltage levels are used in other countries in the world.

The three-phase power supply system has three voltages with equal magnitude, frequency and displaced by 120° each other. That means, the phase angle difference between three phases is 120 electrical degrees. The vector displacement of three-phase voltage is shown in Figure 1.7.

The three-phase voltages are denoted as $e_R$, $e_Y$ $e_B$, and their expression are given below

$$e_R = E_m \sin(\omega t)$$
$$e_Y = E_m \sin(\omega t - 120°)$$
$$e_B = E_m \sin(\omega t + 120°) = E_m \sin(\omega t - 240°)$$

As Phasor rotates in anti-clockwise direction, voltage $e_Y$ lags $e_R$ by 120° and $e_B$ lags $e_R$ by 240°.

The phase angle displacement of three-phase voltage wave shape is shown in Figure 1.8.

The voltage displacement between the phases should be 120 electrical degrees. 360 electrical degrees equal to 20.066 milli seconds from Figure 1.8. Corresponding electrical degree for 6.793 milli seconds is worked out as 121.8 electrical degrees. Figure 1.9 shows the angular separation between three phases. Y phase voltage lags R phase by 120.938° similarly B phase lags Y phase 120.938°. Figure 1.8 and Figure 1.9 represent the phase angle displacement in time domain and angular domain form, respectively.

Line to neutral voltage is voltage between any one of the line to neutral. This voltage is applicable where neutral conductor is also a return current carrying conductor in the circuit like in three-phase four-wire circuit and single-phase two-wire circuit. Low voltage distribution system uses the

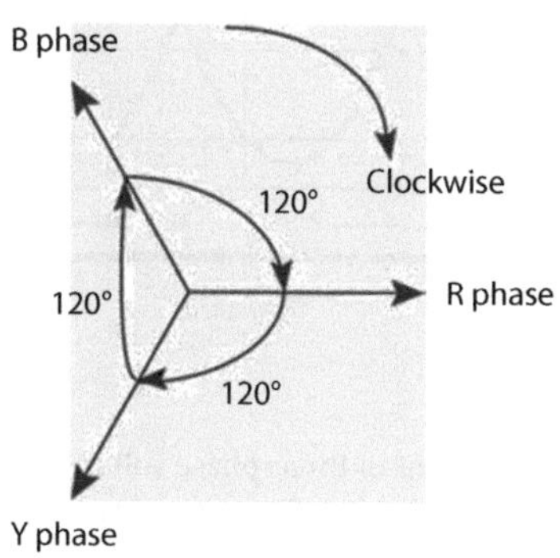


**Figure 1.7** Vector displacement of three phases.

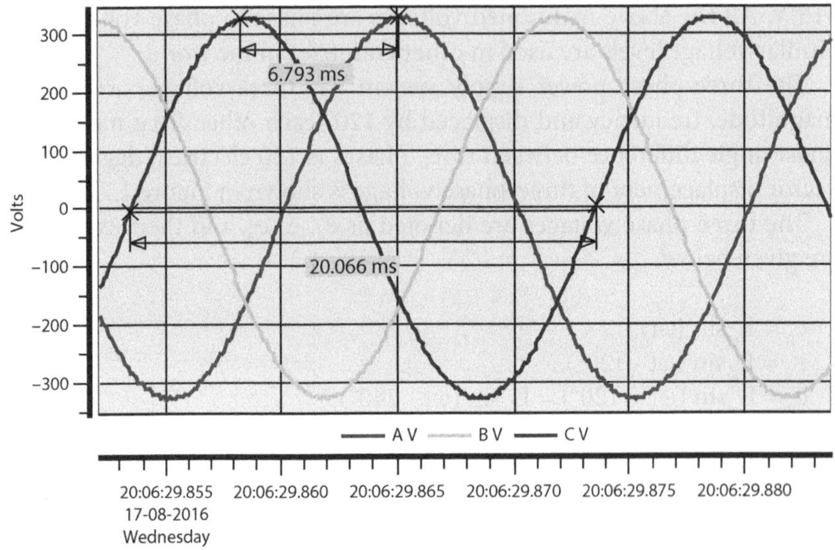


**Figure 1.8** Phase angle displacement of three-phase voltage waveform in time domain. Note: This figure is captured using Dranetz Power Quality analyser.

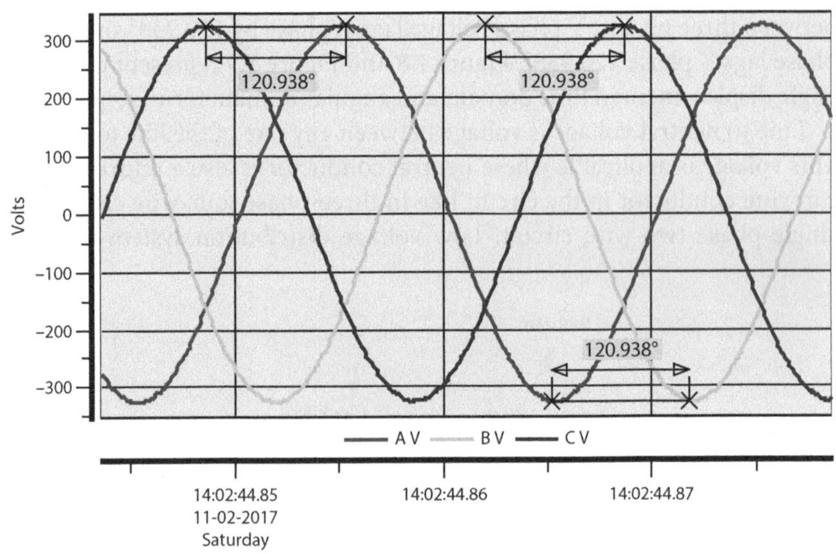


**Figure 1.9** Phase angle displacement of three-phase voltage waveform in angular form. Note: This figure is captured using Dranetz Power Quality analyser.

neutral for single-phase loads. The three-phase, four-wire, 415 V, 50 Hz AC circuit is shown in Figure 1.10.

**Example 1.9:** What is the voltage between R phase and neutral in Figure 1.10?
**Answer:** from eqn. 1.4, R phase and neutral voltage is 240 V.

The single-phase, two-wire AC circuit is shown in Figure 1.11.

**Example 1.10:** What is the voltage between R phase and neutral in Figure 1.11?
**Answer:** 240 V.

### *1.1.1.2 Current*

Current is rate of flow of electric charge across the potential difference in a closed electric circuit. The current flows from high voltage to low voltage in the closed circuit [1]. The SI unit of current is Ampere (A). Higher values of current are mentioned as kA.

Andre Marie Ampere (20th Jan 1775 – 10th Jun 1836): A French scientist discovered the science of classical electromagnetism and electrodynamics. In order to honour him, SI unit of electric current is his name, Ampere.

Andre Marie Ampere. Courtesy: Google image.

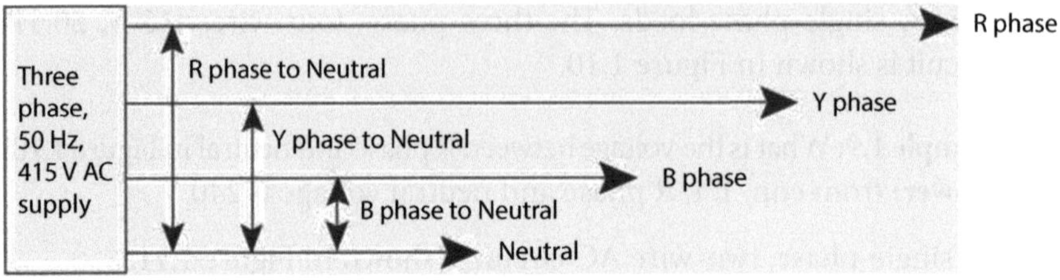


**Figure 1.10** Three-phase, four-wire circuit configuration.

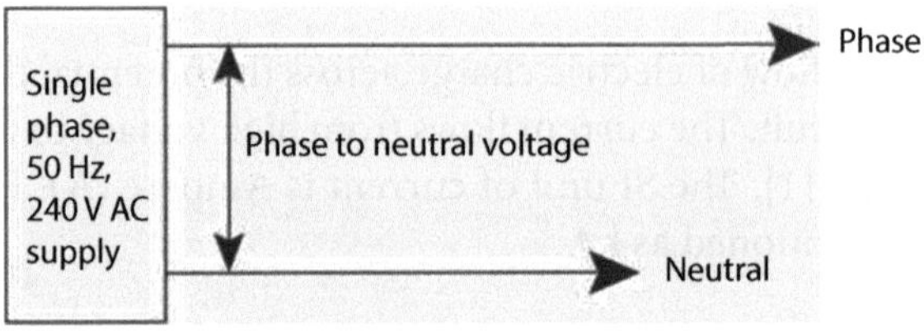


**Figure 1.11** Single-phase, two-wire system.

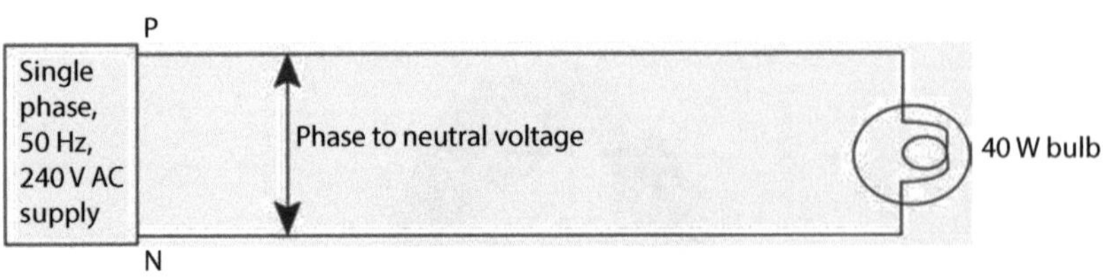


**Figure 1.12** 40 W bulb connected across 240 V supply.

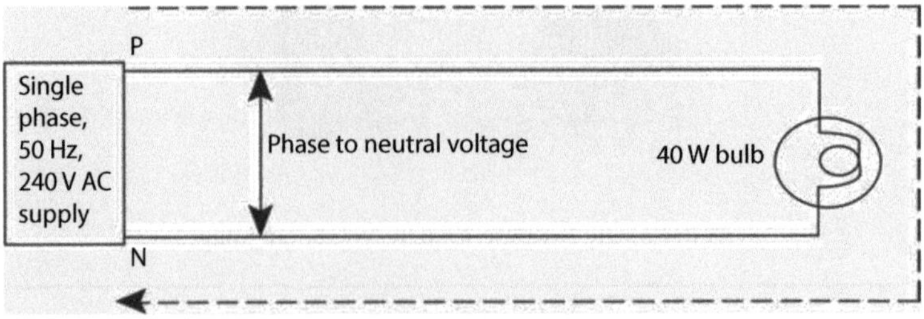


**Figure 1.13** Direction of current flow for positive half cycle.

A single-phase, two-wire AC circuit is shown in Figure 1.12.

The current flow direction for positive half cycle is represented as dotted line in Figure 1.13.

The current flow direction for negative half cycle is represented as dotted line in Figure 1.14.

The factor which decides the value of current flow in the circuit is load impedance.

**Example 1.11:** A 5 Ω resistance (load) is connected across a single-phase 230 V, 50 Hz AC supply is shown in Figure 1.15. Calculate the current drawn by 5Ω resistance.

**Solution:**

Theoretical calculation: Current drawn by the load resistance of 5 Ω is theoretically calculated by Ohms law [31].

$$V = IR \tag{1.5}$$

$$I = \frac{V}{R} \tag{1.6}$$

$$I = 230/5$$

$$I = 46\ A$$

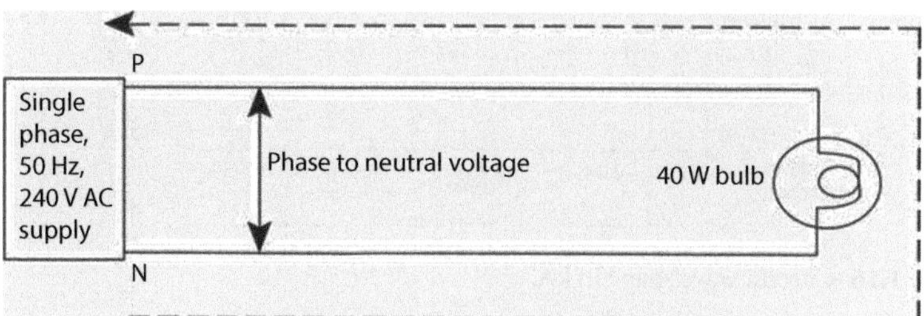


**Figure 1.14** Direction of current flow for negative half cycle.

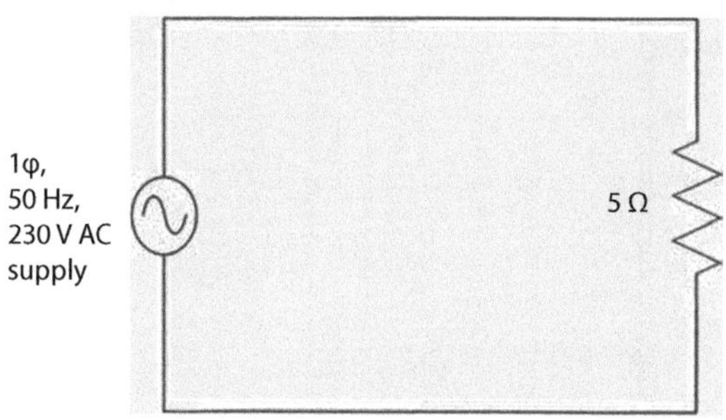


**Figure 1.15** Circuit diagram of single-phase AC supply feeding R Load.

Simulation result:

The instantaneous current (IL) wave shape drawn by the load and RMS current trend are shown in Figure 1.16 and Figure 1.17, respectively.

The current in RMS value of a Sine wave from simulation is 0.046 kA or 46 A. i.e,

$$I_{RMS} = \frac{I_{instantaneous}}{\sqrt{2}} = 46 \text{ A}$$

> George Simon Ohm (16th March 1789 – 6th July 1854): A German scientist discovered direct proportionality between the potential difference applied across a conductor and the resultant electric current. This relationship is called Ohm's law. In order to honour him, SI unit of electric resistance is in his name. Ohm.

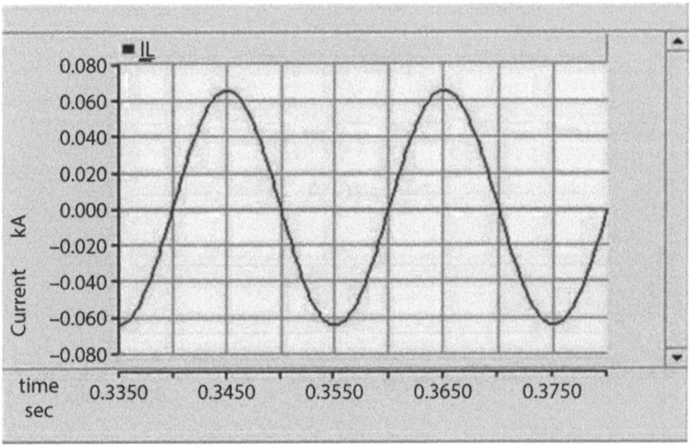


**Figure 1.16** Current wave shape in kA.

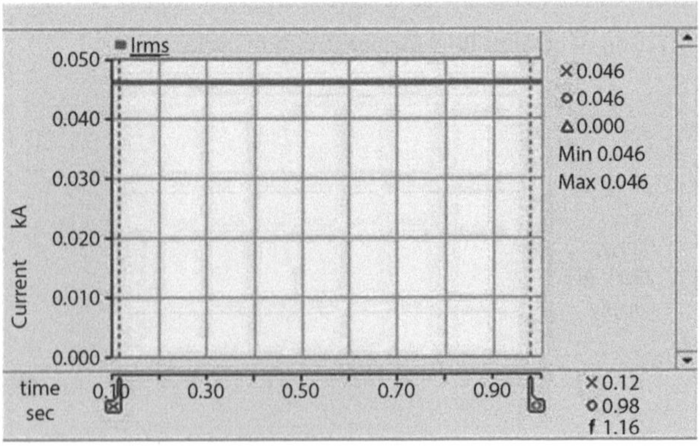


**Figure 1.17** RMS current trend in kA.

### 1.1.1.3 Time Period and Frequency

Time period is the duration of one full cycle equalling one positive half cycle and one negative half cycle for a span of 360°. Frequency is the physical count of positive and negative half cycles appearing in one second. In India, the power supply frequency is 50 Hz. The 50 Hz frequency has 50 numbers of positive half cycle and 50 numbers of negative half cycle in one second. In the United States, the power supply frequency is 60 Hz. The 60 Hz frequency has 60 numbers of positive half cycle and 60 numbers of negative half cycle in one second. Conversely, one upon frequency is the time period. The time period of one cycle is 20 milli second for 50 Hz frequency.

$$F = \frac{1}{T} \tag{1.7}$$

$$T = \frac{1}{F} \tag{1.8}$$

Figure 1.18 shows the wavelength of 19.994 milli seconds which corresponds to one cycle. Figure 1.19 shows 50.02 cycles appear in one second.

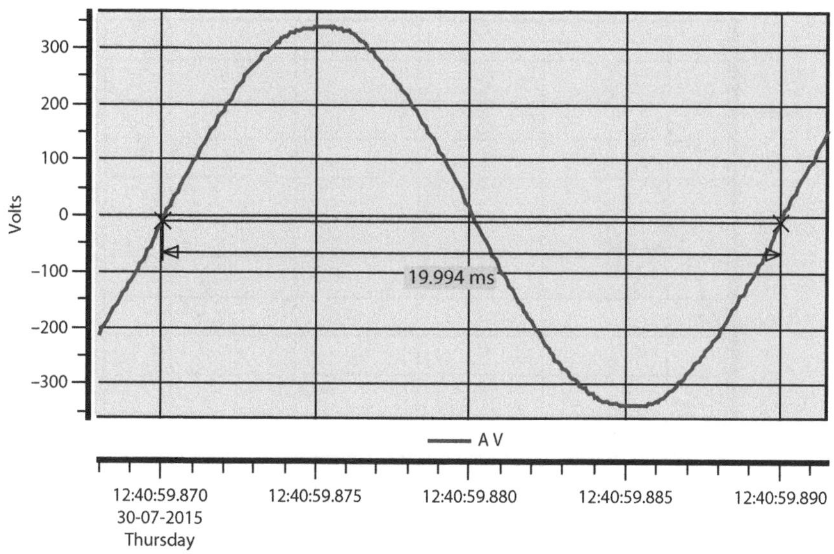


**Figure 1.18** Time period. Note: This figure is captured using Dranetz Power Quality analyser.

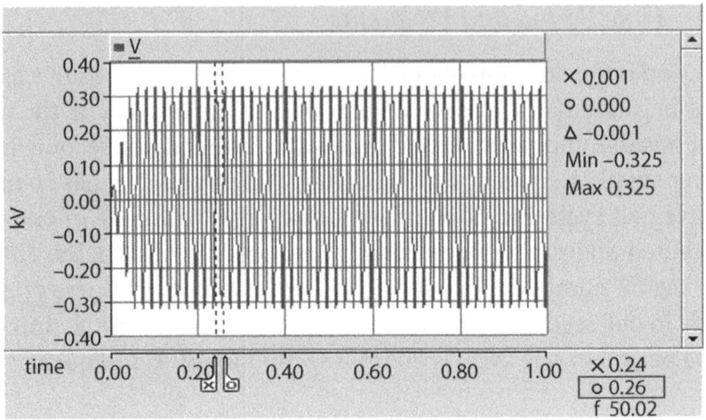


**Figure 1.19** Voltage frequency.

In other words, one upon 19.994 in Figure 1.18 equals 50.02 which is reflected in Figure 1.19.

### 1.1.1.4 *Phase Angle* ($\phi$)

The angular displacement between the voltage and current wave shapes determine whether current leads the voltage or current lags the voltage

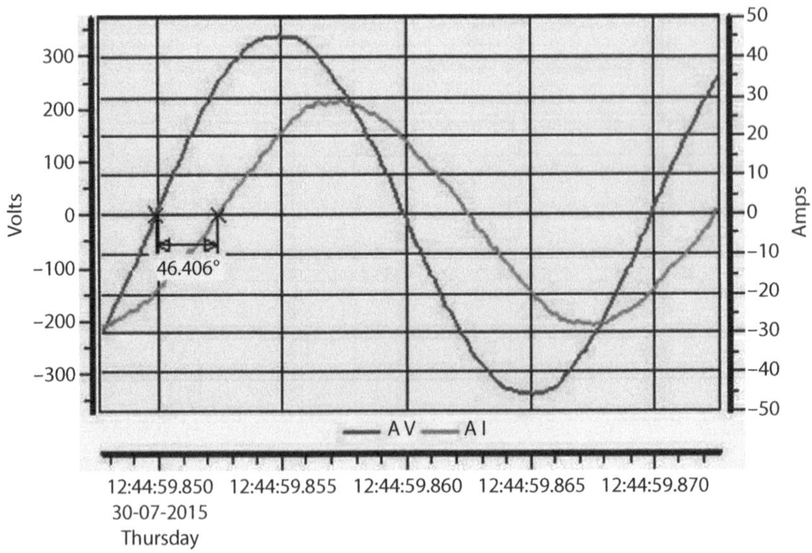


**Figure 1.20** Phase angle between voltage and current. Note: This figure is captured using Dranetz Power Quality analyser.

or both are in phase. This factor is vital in AC system as this will decide whether the product of voltage and current is positive or negative. The phase displacement in angular degrees between voltage and current wave shape is 46.406° as shown in Figure 1.20

The cosine value of the phase angle displacement is called as displacement power factor without presence of multiple frequencies (harmonics) introduced by modern power electronics devices explained in detail in section 1.1.4.

## 1.2 Three-Phase Supply Connections

In a single-phase system, two wires (line and neutral) are functionally sufficient for transferring the power to the load. The three-phase system, three wires (R, Y, B) or four wires (R, Y, B, N) are functionally required for transferring the power to the loads. Three wires are required for powering the three-phase loads without neutral like motor loads and four wires required for three-phase loads with neutral like UPS systems [22]. The three-phase power supply connections are classified into two types.

1) Star connection or wye connection
2) Delta connection

### 1.2.1 Star Connection

The star connection is formed by connecting starting or terminating ends of all the three windings of transformer or generator together as shown in Figure 1.21. The one ends of the winding $R_2 - Y_2 - B_2$ are connected together and formed as star connection. Other ends of the winding $R_1 - Y_1 - B_1$ are connected to the loads. The common point N is called Neutral point. The phases

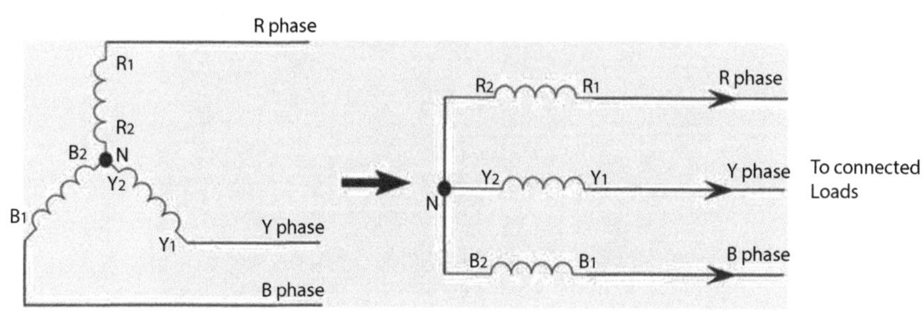


**Figure 1.21** Star circuit connection.

$R_1 - Y_1 - B_1$ and common point N as neutral is extended to the connection points where the different loads are getting connected. The star circuit is suitable for six numbers of possible loads combination such as RY, YB, BR, RN, YN and BN.

The star connection is used in three-phase AC generator and transformers [6]. The name plate details of three-phase AC generator and transformer are shown in Figure 1.22 and Figure 1.23, respectively.

The terminal connection of transformer star winding in actual site installation is shown in Figure 1.24.

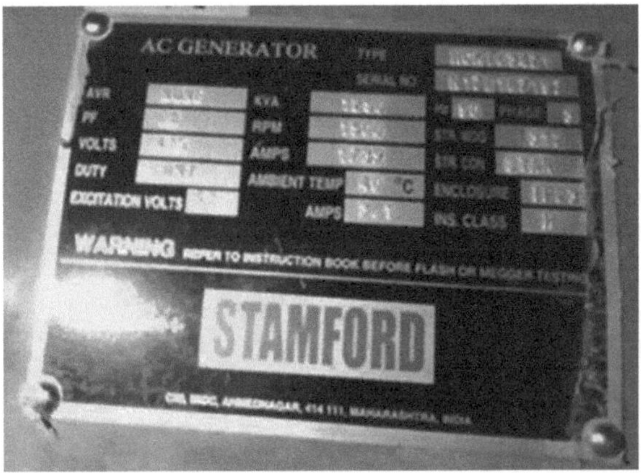


**Figure 1.22** Name plate details of AC generator (Courtesy: Stamford).

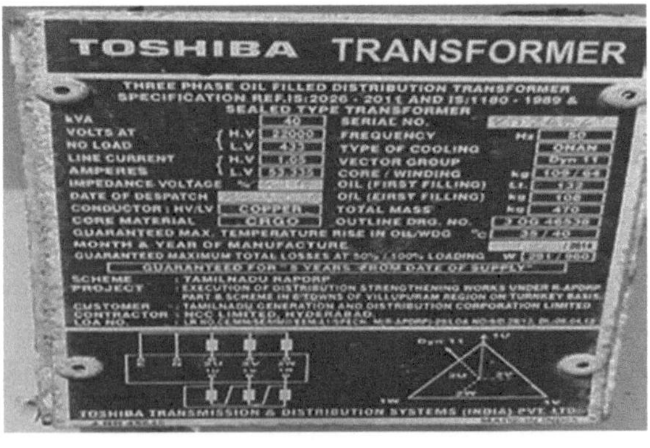


**Figure 1.23** Name plate details of transformer (Courtesy: Toshiba).

**Figure 1.24** Terminal connection of transformer secondary side - star winding.

**Advantages of star circuit:**

The star circuit has the following advantages over delta circuits:

1) The line voltage in star circuit is phase voltage multiplied by $\sqrt{3}$ and current drawn by star circuit is less as compared with current drawn by delta circuit. Hence, required conductor size for star circuit is lesser for same rating.
2) Star circuit has the neutral point, hence three-phase, four-wire system can be developed.
3) It provides the two different voltage levels for the load connection. For an example, 415 **V** between line to line and 240 **V** between line to neutral.
4) Both single-phase and three-phase loads can be connected in single star circuits.

**Disadvantages of star circuit:**

The star circuit has the following disadvantages over delta circuits:

1) Higher size neutral conductor is required for highly unbalanced and harmonic producing loads.

### 1.2.2 Delta Connection

The delta connection is formed by connecting one end of the winding to starting point of other winding to form a closed loop as shown in Figure 1.25.

The winding end $R_2$ is connected with starting point of another winding $Y_1$, winding end $Y_2$ is connected with starting point of another winding $B_1$ and winding end $B_2$ is connected with starting point of another winding $R_1$. The output connections are taken at mid of the two windings such as $B_2 - R_1$, $R_2 - Y_1$ and $Y_2 - B_1$. Delta circuits are suitable for three numbers of loads combination such as RY, YB and BR.

The practical connection or forming delta circuit in transformer is shown in Figure 1.26 and Figure 1.27. In Figure 1.26, winding end $R_2 - Y_1$ is combined and output is taken at the combined point. This transformer

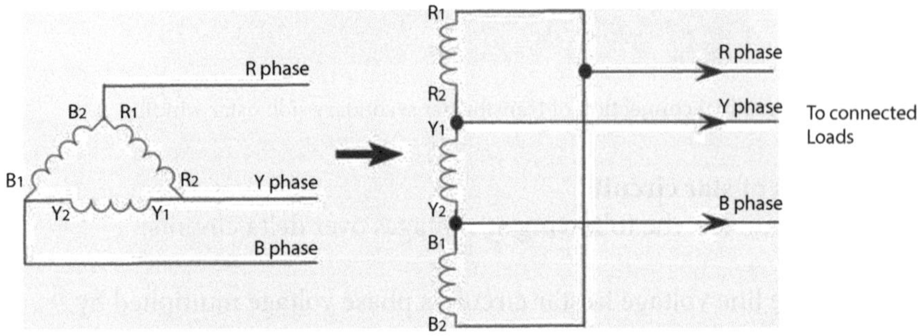


**Figure 1.25** Delta circuit connection.

**Figure 1.26** Practical connection or forming delta circuit in transformer.

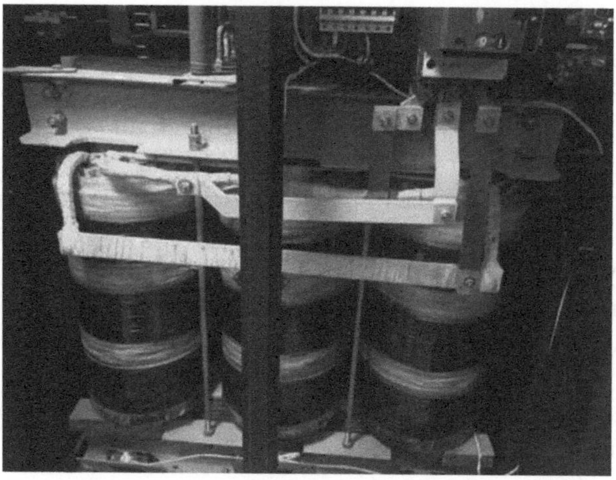

**Figure 1.27** Practical connection or forming delta circuit in transformer.

(1:1 ratio transformer) used at output side of the UPS system for providing the isolation in the circuit (415 V level).

In Figure 1.27, winding end $B_2 - R_1$ is combined and output is taken at the combined point is R phase (Red color), winding end $R_2 - Y_1$ is combined and output is taken at the combined point is Y phase (Yellow color) and winding end $Y_2 - B_1$ is combined and output is taken at the combined point is B phase (Blue color).

**Advantages of delta circuit:**
The delta circuit has the following advantages over star circuits:

1) More suitable for three-phase loads like induction motors
2) Unbalancing is avoided

**Disadvantages of delta circuit:**
The delta circuit has the following disadvantages over star circuits:

1) Suitable for three-phase loads only; connection of single-phase loads are not possible.

### 1.2.3 Balanced Load

The loads are called balanced loads when drawing the balanced or equal current magnitude in all the three phases from the power supply [7]. In other words, impedance connected across the supply is equal for all the three phases. The delta circuit having the impedances $Z_{RY}$, $Z_{RB}$ and $Z_{YB}$ connected across the RY, YB and RB phases respectively i.e $Z_{RY} = Z_{RB} = Z_{YB}$. If the impedance of $Z_{RY}$, $Z_{RB}$ and $Z_{YB}$ is equal/same, then the circuit is called as balanced delta circuit shown in Figure 1.28.

The star circuit having the impedances $Z_R$, $Z_Y$ and $Z_B$ connected across the RY, YB and RB phases respectively and RN, YN and BN line to neutral respectively, i.e., $Z_R = Z_Y = Z_B$. If the impedance of $Z_R$, $Z_Y$ and $Z_B$ is same, then the circuit is called as balanced star circuit and shown in Figure 1.29.

Practically most of the three-phase induction motors have equal impedances in all three phases. The magnitude of current flowing to the motor is almost equal in all the three phases. These three-phase motor loads are called balanced loads.

**Example 1.12:** A three-phase, 11 kW induction motor name plate details is shown in Figure 1.30.

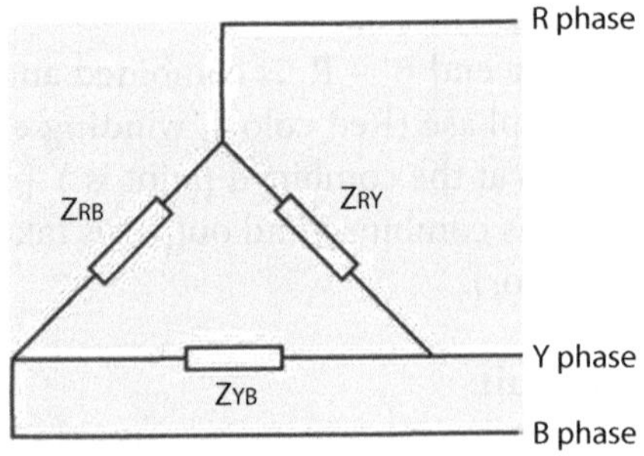


**Figure 1.28** Balanced delta circuit.

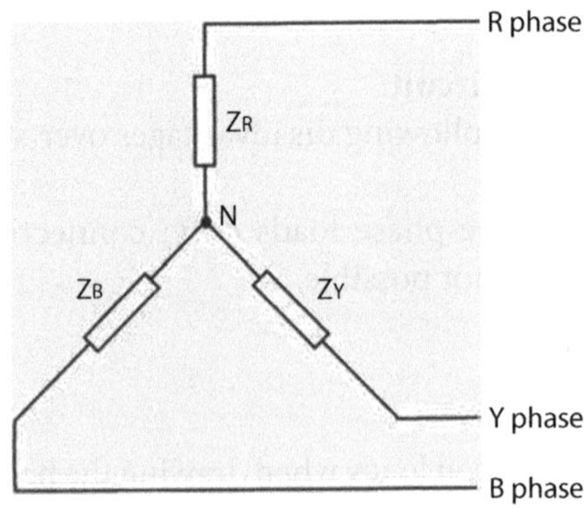


**Figure 1.29** Balanced star circuit.

**Figure 1.30** Name plate details of three phase induction motor (Courtesy: TECO).

When the motor is operating at 415V, 50Hz frequency at 0.88 power factor, it draws the 20.5A current/phase in delta connection and 11.8A current/phase in star connection. The same motor, when operating at 380V, 60Hz frequency at 0.85 power factor, draws the 21A current/phase in delta circuit and 11.6A current/phase in star circuit.

### 1.2.4 Unbalanced Load

Loads are called unbalanced loads when drawing the unbalanced or unequal current magnitude in all the three phases from the supply [7]. In other words, impedance connected across the supply is not equal in all the three phases, i.e., $Z_{RY} \neq Z_{YB} \neq Z_{BR}$. If the impedance of $Z_{RY}$, $Z_{YB}$ and $Z_{RB}$ is not equal, then the circuit is called as unbalanced circuit.

### 1.2.5 Star – Delta Conversion

The Star-Delta conversion is used to convert the impedances in star circuit to equivalent impedance in delta circuit. The star-delta conversion is shown in Figure 1.31.

In general, resistance of delta circuit is calculated from eqn 1.9.

$$R_{\Delta} = \frac{R_P}{R_{opposite}} \tag{1.9}$$

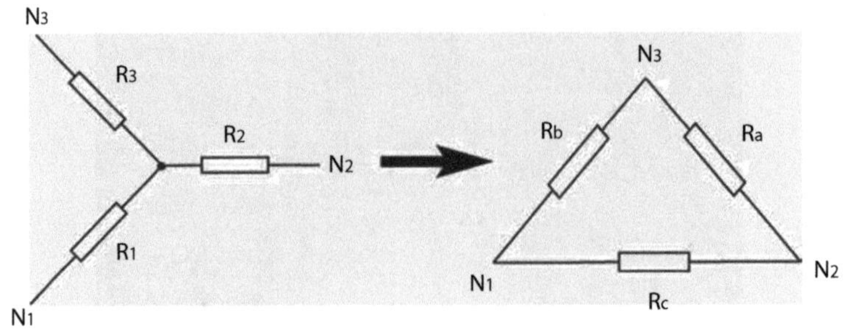


**Figure 1.31** Star to delta conversion.

Where

$R_P = R_1R_2 + R_2R_3 + R_3R_1$

From eqn (1.9), the resistance $R_a$, $R_b$. $R_c$ is calculated as

$$R_a = \frac{R_1R_2 + R_2R_3 + R_3R_1}{R_1}$$

$$R_b = \frac{R_1R_2 + R_2R_3 + R_3R_1}{R_2}$$

$$R_c = \frac{R_1R_2 + R_2R_3 + R_3R_1}{R_3}$$

$R_a$, $R_b$ and $R_c$ is delta circuit impedances
$R_1$, $R_2$ and $R_3$ is star circuit impedances

### 1.2.6 Delta to Star Conversion

The Delta-Star conversion is used to convert the impedances in delta circuit to equivalent impedance of star circuit. The delta-star conversion is shown in Figure 1.32.

$$R_1 = \frac{R_bR_c}{R_a + R_b + R_c}$$

$$R_2 = \frac{R_aR_c}{R_a + R_b + R_c}$$

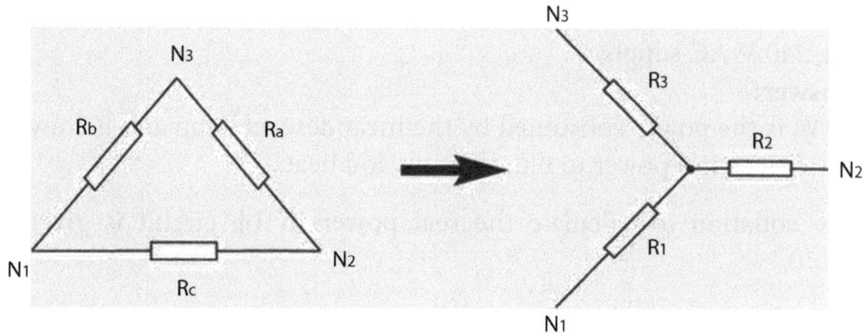


**Figure 1.32** Delta to star conversion.

$$R_3 = \frac{R_a R_b}{R_a + R_b + R_c}$$

Where

$R_a$, $R_b$ and $R_c$ is delta circuit impedances
$R_1$, $R_2$ and $R_3$ is star circuit impedances

## 1.3 Power

Electricity is fed to devices which in turn do the work for us. For example, an electric heater delivers thermal power (heat) and a motor delivers mechanical power. Both the devices consume electrical power and deliver different forms of output which are utilized directly. Electrical power is a product of voltage and current. Depending upon the type of circuit, electrical power is classified into three forms:

1) Real power (W)
2) Reactive power (VAr)
3) Apparent power (VA)

### 1.3.1 Real Power or Active Power (P)

Real power is the power that gets consumed by the load to deliver useful output. The SI unit of real power is expressed in Watts (W). Higher values of real power are mentioned as kW, MW and GW.

**Example 1.13:** A 40 W incandescent lamp is connected across single-phase, 240 V AC supply.

**Answer:**

40 W is the power consumed by the incandescent lamp and it converts 40 W of electrical power to illumination and heat.

The equation to calculate the real power in 1ϕ circuit is given in eqn 1.10.

$$P = VI \cos(\phi) \tag{1.10}$$

Where

P is real power in W

V is line to neutral voltage in V

I is current in A

Cos ϕ is power factor

**Example 1.14:** A 5Ω resistive load is connected across single-phase, 240 V AC supply as shown in Figure 1.33. Power factor is unity. Calculate the real power consumed by the resistive load.

**Answer:**

Load resistance is 5Ω

Power factor is unity

Current flow from eqn 1.6, 240/5 = 48 A

Real power consumed by the load from eqn 1.10, 240 x 48 x 1 = 11520 W

Real power consumed by the resistive load is 11520 W or 11.52 kW.

The equation to calculate the real power in 3ϕ balanced circuit is

$$P = \sqrt{3} VI \cos(\phi) \tag{1.11}$$

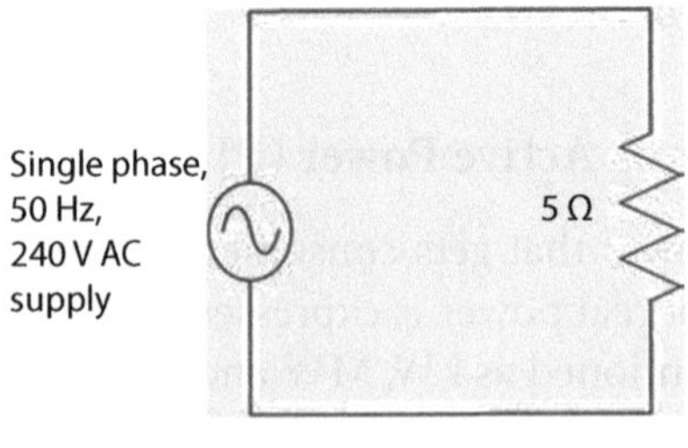


**Figure 1.33** Single phase, 240V circuit powering resistive (5 Ω) load.

Where

P is real power in W

V is phase to phase voltage in V

I is per phase current in A

Cos $\phi$ is power factor

**Example 1.15:** A 5Ω resistive load is connected across the all phases as shown in Figure 1.34. Circuit is assumed to be balanced and power factor is unity. Calculate the real power consumed by the resistive load.

**Answer:**

Voltage is 415 V

Load resistance is 5Ω/ph

Power factor is unity

From eqn. 1.4, line to neutral voltage is 240 V.

Current flow per phase from eqn 1.6, 240/5 = 48 A

Real power consumed by the load from eqn 1.11, = 1.732 x 415 x 48

Total power consumed by the load is 34501 W or 34.5 kW.

For an unbalanced circuit, real power in eqn 1.11 is not applicable and the expression to calculate the real power in 3ϕ unbalanced circuit is given in 1.12.

$$P = V_r I_r \cos(\phi_r) + V_y I_y \cos(\phi_y) + V_b I_b \cos(\phi_b) \qquad (1.12)$$

Where

P is real power in W

$V_r$, $V_y$, $V_b$ are the respective line to neutral voltages in V

$I_r$, $I_y$, $I_b$ are the respective phase currents in A

Cos $\phi_r$, Cos $\phi_y$, Cos $\phi_b$ are the respective power factors

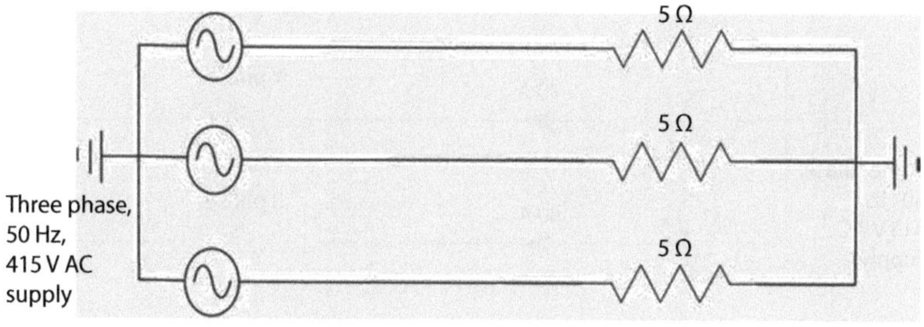


**Figure 1.34** Three-phase, 415V balanced circuit powering the resistive load (5 Ω/phase).

**Example 1.16:** A 415 V distribution system has unbalanced loads across the three phases. Calculate the real power for phase R has 50A at 0.8 lagging power factor, phase Y has 70A at 0.9 lagging power factor and phase B has 80A at unity power factor as shown in Figure 1.35.

**Answer:**

From eqn 1.4, $V_{ph} = \frac{V_L}{\sqrt{3}} = \frac{415}{\sqrt{3}} = 240V$ (line to neutral voltage is 240 V).

$V_r = V_y = V_b = 240$ V.

$I_r = 50A$, $I_y = 70A$, $I_b = 80A$

$Cos\phi_r = 0.8$, $Cos\phi_y = 0.9$, $Cos\phi_b = 1$

Real power consumed by the load from eqn 1.12,

P = (240x50x0.8) + (240x70x0.9) + (240x80x1)

P = 9600 + 15120 + 19200 = 43920 W or 43.92 kW.

### 1.3.2 Reactive Power (Q)

When an AC circuit is energized, a magnetic field is created across the conductor due to reactance offered by the path. The magnetising current drawn by the reactive load to produce the magnetic field helps in consuming active power. The product of applied voltage and magnetising current is the power which is required to consume the active power. This power is known as reactive power, which is essential for active power consumption [8]. In other words the power drawn by the reactive component of the load is reactive power.

The reactive power is expressed as Voltage Ampere reactive (VAr). Higher values of reactive power are mentioned as kVAR and MVAR.

The equation for calculate the reactive power in single-phase circuit is

$$Q = VI \sin(\phi) \tag{1.13}$$

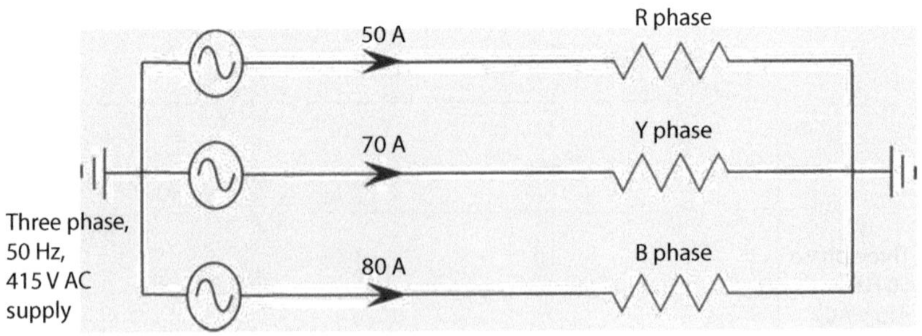


**Figure 1.35** Three-phase, 415V unbalanced circuit.

Where

Q is reactive power in VAR
V is line to neutral voltage in V
I is phase current in A
ϕ is angle between voltage and current

**Example 1.17:** A 5Ω load at 0.8 PF connected across single phase, 240 V AC supply as shown in Figure 1.36. Calculate the reactive power drawn by the load.

**Answer:**

Load is 5Ω
Power factor is 0.8
From eqn 1.6, 240/5 = 48 A
$\phi = \text{Cos}^{-1}(0.8) = 36.87°$
Reactive power drawn by the load from eqn 1.13,
Q = 240 x 48 x Sin (36.87°)
= 6912VAr or 6.91 kVAr

The equation for calculate the reactive power in three-phase balanced circuit is

$$Q = \sqrt{3}VI\sin(\phi) \tag{1.14}$$

Where

Q is reactive power in VAr
V is phase to phase voltage in V
I is per phase current in A
ϕ is angle between voltage and current

**Figure 1.36** Single phase, 240V circuit.

**Example 1.18:** A three-phase, 415V, system has 10Ω impedance with 0.85 lagging power factor in each phase as shown in Figure 1.37. Calculate the reactive power drawn by the load.

**Answer:**

Voltage is 415 V

Load resistance is 10 Ω/ph

Power factor is 0.85

From eqn 1.6, current flow is 240/10 = 24 A

$\phi = \text{Cos}^{-1}(0.85) = 31.79°$

Reactive power drawn by the load from eqn 1.14,

$$Q = \sqrt{3} \text{ x } 415 \text{ x } 24 \text{ x } \text{Sin}(31.79°)$$

$$= 9087.8 \text{ VAr or } 9.09 \text{ kVAr.}$$

The equation to calculate the reactive power in three phase, unbalanced circuit is

$$Q = V_r I_r \sin(\phi_r) + V_y I_y \sin(\phi_y) + V_b I_b \sin(\phi_b) \tag{1.15}$$

Where

Q is reactive power in VAr

$V_r, V_y, V_b$ are the respective line to neutral voltages in V

$I_r, I_y, I_b$ are the respective phase currents in A

$\phi_r, \phi_y, \phi_b$ are the angular displacement between voltage and current for the respective phases

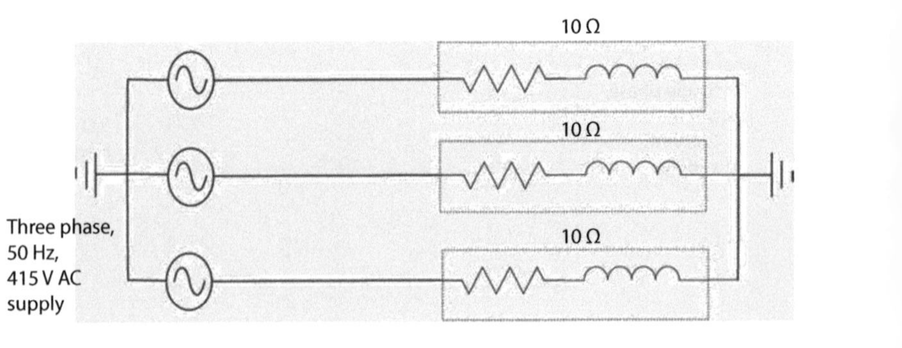


**Figure 1.37** Three-phase, 415V balanced circuit.

**Example 1.19:** A 415 V distribution system has unbalanced loads across the three phases. Calculate the reactive power for phase R has 80A at 0.85 lagging power factor, phase Y has 90A at 0.7 lagging power factor and phase B has 75A at unity power factor as shown in Figure 1.38.

**Answer:**

From eqn. 1.6, line to neutral voltage is 240 V.

$V_r = V_y = V_b = 240$ V.

$I_r = 80A, I_y = 90A, I_b = 75A$

$\cos \phi_r = 0.85, \cos \phi_y = 0.7, \cos \phi_b = 1$

$\phi_r = 31.79°, \phi_y = 45.57°, \phi_b = 0°,$

Reactive power drawn by the load from eqn 1.15,

Q = (240 x 80x sin (31.79°)) + (240 x 90 x sin (45.57°)) + (240 x 75 x sin (0°))

Q = 10114.7 + 15424.7 + 0 = 25539.4 VAr or 25.53 kVAr.

### 1.3.3 Apparent Power (S)

Apparent power is the vector sum of real and reactive power. The apparent power is expressed in Volt Amperes (VA). Vector addition technically means taking the square root of the sum of the square of real power and the square of reactive power. The expression for apparent power is given in equation (1.16).

$$VA = \sqrt{W^2 + VAr^2} \tag{1.16}$$

All the three forms of power are represented using power triangle where real power is represented along the abscissa, reactive power along the ordinate and apparent along the hypotenuse. The power triangle is shown in Figure 1.39.

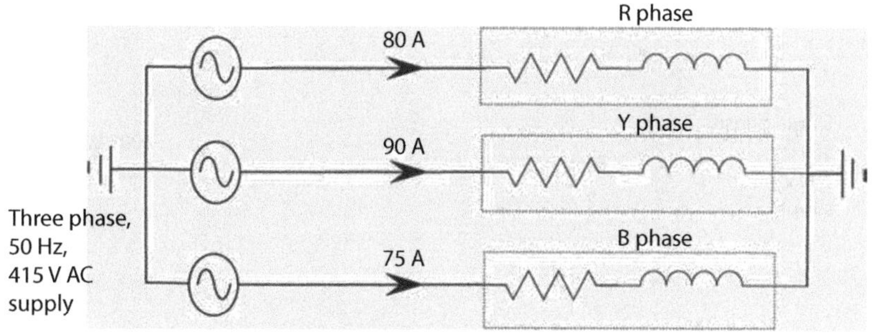


**Figure 1.38** Three phase, 415V unbalanced circuit.

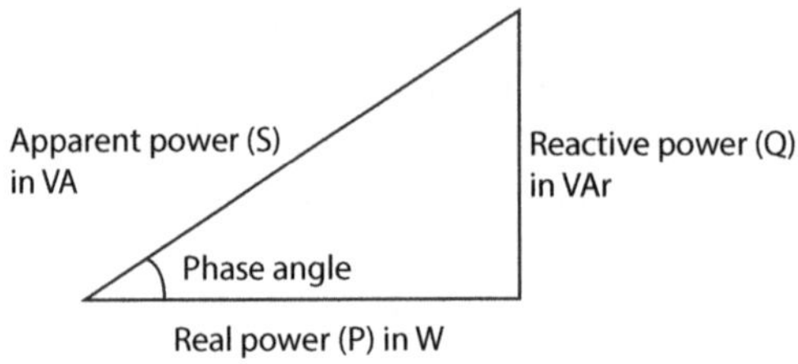


**Figure 1.39** Power triangle.

The equation 1.17 is used for calculating the apparent power in single-phase circuit when voltage and current are known.

$$S = V \text{ x } I \tag{1.17}$$

Where
S is apparent power in VA
V is line to neutral voltage in V
I is current in A

**Example 1.20:** A single-phase, 1000W focus lamp draws 4.6A current in a 240 V AC supply as shown in Figure 1.40. Calculate the apparent power.
**Answer:**
From eqn. 1.17, S = 240 x 4.6 = 1104 VA or 1.1 kVA.

The equation 1.18 is used for calculating the apparent power in three-phase, balanced circuit when voltage and current are known.

$$S = \sqrt{3}VI \tag{1.18}$$

**Figure 1.40** Single-phase circuit powering 1000 W focus lamp.

Where
S is apparent power in VA
V is phase to phase voltage in V
I is per phase current in A

**Example 1.21:** A three-phase, 415V system has 10Ω load impedance with 0.85 lagging power factor in each phase as shown in Figure 1.41. Calculate the apparent power in the circuit.

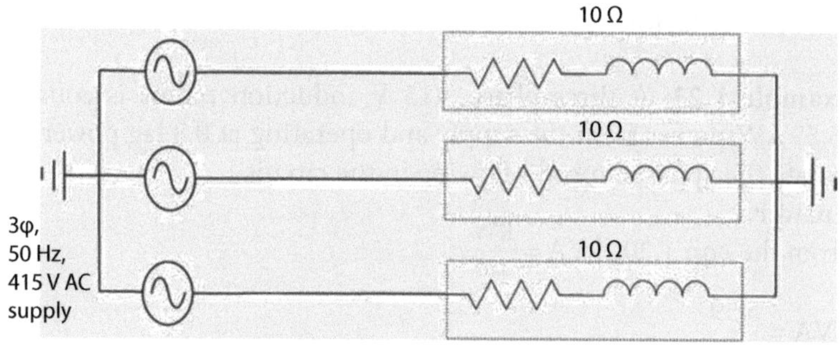


**Figure 1.41** Three-phase, balanced circuit.

**Answer:**
From eqn. 1.4, line to neutral voltage is 240 V.
Per phase current from eqn 1.6, 240/10 = 24 A
$S = \sqrt{3} \text{ x } 415 \text{ x } 24$
S = 17251.23 VA or 17.25 kVA

The expression for calculating the apparent power from real power and reactive power is given in eqn. 1.19.

$$VA = \sqrt{W^2 + VAr^2} \tag{1.19}$$

Where
VA is apparent power
W is real power
VAr is reactive power

**Example 1.22:** A load draws 2.3 kVAr reactive power to consume 5.59 kW of real power. Calculate the apparent power.
**Answer:**
kW = 5.59
kVAr = 2.3
From eqn 1.19, $kVA = \sqrt{kW^2 + kVAr^2}$
$kVA = \sqrt{5.59^2 + 2.3^2}$
$= 6.04$ kVA

The expression for calculation of apparent power from real power and power factor is given in eqn 1.20.

$$VA = \frac{W}{PF} \tag{1.20}$$

Where

VA is apparent power
W is real power
PF is power factor

**Example 1.23:** A three-phase, 415 V induction motor is consuming the 5.59 kW power from the supply and operating at 0.8 lag power factor. Calculate the apparent power flowing in the circuit.

**Answer:**

From the eqn 1.20, $kVA = \frac{kW}{PF}$

$$kVA = \frac{5.59}{0.8}$$

Apparent power = 6.99 kVA

The expression for calculating the apparent power in three-phase, unbalanced circuit is given in eqn 1.21.

$$S = V_r I_r + V_y I_y + V_b I_b \tag{1.21}$$

Where

S is apparent power in VA

$V_r$, $V_y$, $V_b$ are the respective line to neutral voltages of R, Y and B phases respectively and expressed in Volts

$I_r$, $I_y$, $I_b$ are the respective phase currents of R, Y and B phases respectively and expressed in Amps

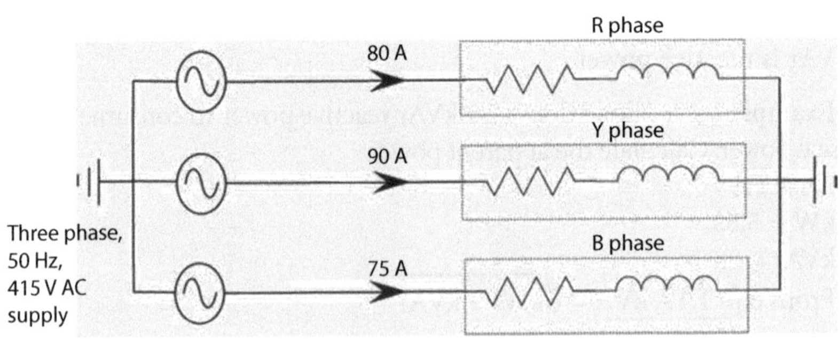


**Figure 1.42** Three phase, 415V unbalanced circuit.

**Example 1.24:** A 415 V distribution system has unbalanced loads across three phases. Phase current of 80A, 90A and 75A flow in R, Y and B phases, respectively, as shown in Figure 1.42. Calculate the real power when R phase was at 0.85 power factor, Y phase was at 0.7 PF and B phase at unity power factor.

**Answer:**

From eqn. 1.4, line to neutral voltage is 240 V.

$V_r = V_y = V_b = 240$ V.

$I_r = 80A, I_y = 90A, I_b = 75A$

$S = (240 \times 80) + (240 \times 90) + (240 \times 75)$

Apparent power = 19200 + 21600 + 18000 = 58800 VA or 58.8 kVA.

## 1.4 Power Factor (PF)

Power factor is the effective usage of real power converted into useful work from apparent power. PF is a load dependent parameter and it comes into the picture for inductive and capacitive loads. When current flows in inductive circuit, reactive power is used for magnetization. Utilization of more reactive power to consume active power reduces the power factor [8, 11]. In other words, power factor determines how effectively real power is consumed in the electric circuit. The expression for true power factor is given in eqn. 1.22.

$$PF = \frac{\text{Real Power (W)}}{\text{Apparent Power (VA)}} \tag{1.22}$$

The power factor is generally classified into two types

i) Classification based on R, L and C through load characteristics

ii) Classification based on harmonics producing loads

### 1.4.1 Classification Based on Load Characteristics

Power factor is defined as the ratio of real power (W or kW) to apparent power (VA or kVA) and expression for power factor is given in eqn 1.22. The power factor is unity when phase angle difference between voltage and current is 0°. The power factor is zero when phase angle difference between voltage and current is 90°. Hence, the value of power factor lies between 0 and 1.

Where

1 is highest power factor when real power equals apparent power or angular displacement between voltage and current is 0°

0 is lowest power factor when real power is zero and reactive power equals apparent power or angular displacement between voltage and current is 90°.

The power factors are classified into three types based on linear relation between voltage and current characteristics:

- ➢ Unity power factor
- ➢ Lagging power factor
- ➢ Leading power factor

The relation between voltage and current for pure resistive, inductive and capacitive loads are listed in Table 1.1.

**Unity power factor (Power factor value is 1):**

The unity power factor is where the current follows the voltage in phase or angular displacement between voltage and current is zero. The following loads are resistive in nature and draw unity power factor from the supply:

- ➢ Incandescent lamp
- ➢ Water heater
- ➢ Soldering iron
- ➢ Iron box

**Example 1.25:** Unity PF representation using PSCAD simulation

A single phase 230 V, 50 Hz AC supply powers a resistive load of 2Ω. Circuit diagram of ideal resistive load connected to AC supply is shown in Figure 1.43.

Figure 1.44 shows the relationship between voltage and current wave shape for unity power factor load. The connected load resistance of 2Ω draws the peak current from the source is sqrt (2) x V/R = 162 A is shown in Figure 1.44.

**Example 1.26:** Unity power factor using practical measurement

A resistive load bank of 190 kW is connected in the output side of UPS system whose name plate details are 500 kVA at 0.8 PF, output voltage is

**Table 1.1** Voltage and current relation for resistive, inductive and capacitive loads.

| Type of load | Relation of voltage and current | Power factor | Waveform |
|---|---|---|---|
| Resistor | Current in phase with voltage | Unity | 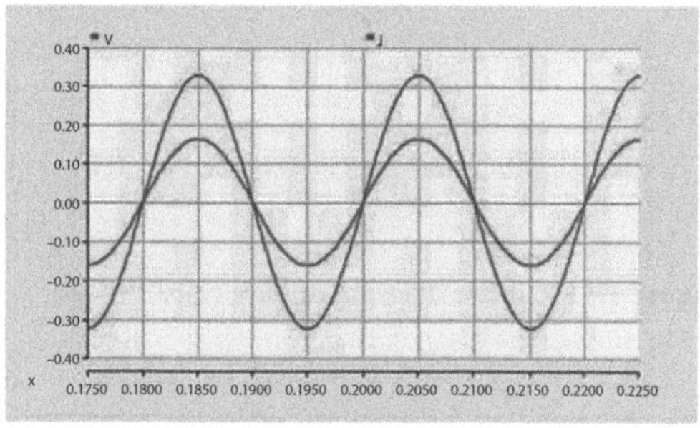  |
| Inductor | Current lags the voltage by 90° | Lagging | 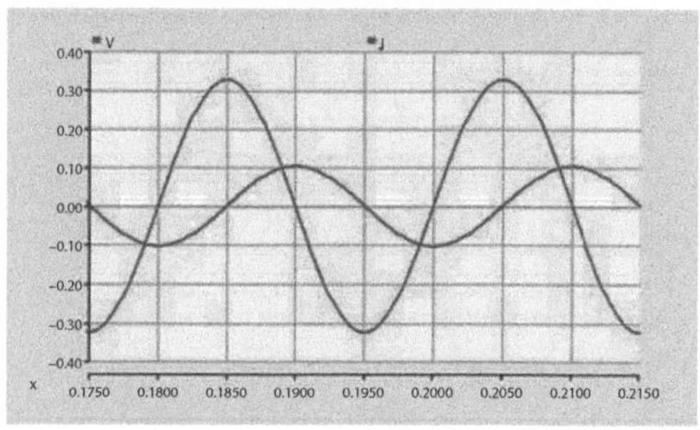  |
| Capacitor | Current leads the voltage by 90° | Leading | 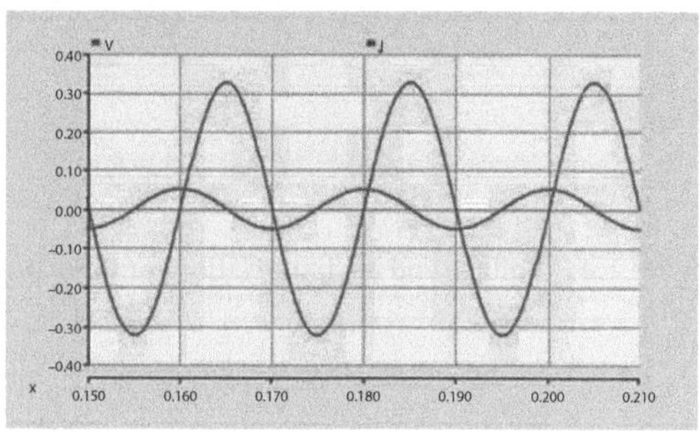  |

400V, 50 Hz, three phase, four wire. The schematic diagram of the power distribution is shown in Figure 1.45. Power quality analyzer is used to monitor the power parameters at the output side of the UPS system for the duration of one minute and 50 seconds between 12:36:00 to 12:37:50 hours.

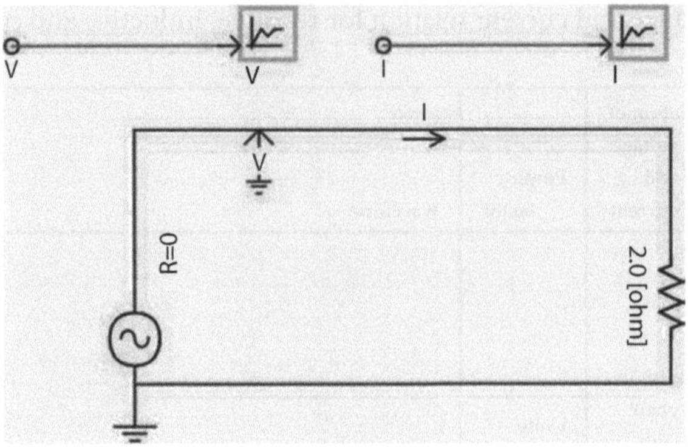


**Figure 1.43** Circuit diagram of pure resistive load.

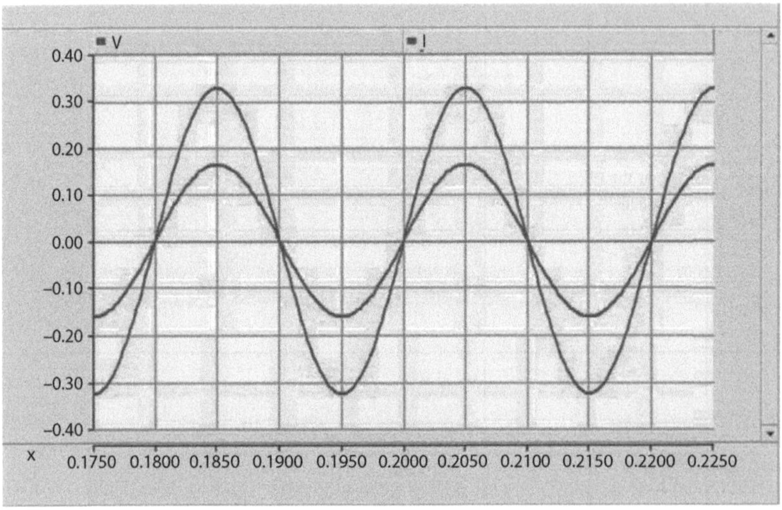


**Figure 1.44** Voltage and current relation for unity power factor load.

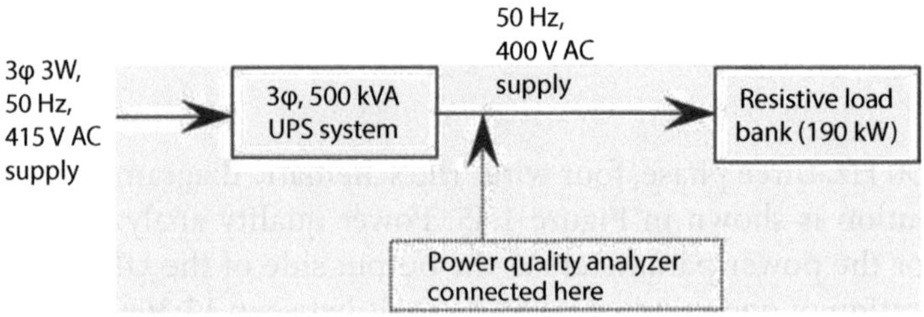


**Figure 1.45** Schematic diagram.

Figure 1.46 shows the relationship between voltage and current characteristics drawn by resistive load.

Figure 1.47 shows the relationship between real power, apparent power and PF trend. The ratio of real power (69.97 kW) to apparent power (69.97 kVA) is equal to 1. That means, the resistive loads draws the real power from the UPS system at unity power factor.

**Lagging power factor (Power factor value is between 0 and 0.99):**
The lagging power factor is the angular displacement of voltage and current greater than 0° but less than or equal to 90°. Current wave shape drawn by an ideal inductor lags the voltage wave shape by 90°. The lagging power factor is generally denoted by a '+' sign. The following loads are inductive in nature and draw the real power at lagging power factor.

- Transformers
- Motors
- Induction furnace

**Example 1.27:** Lagging power factor representation using PSCAD simulation

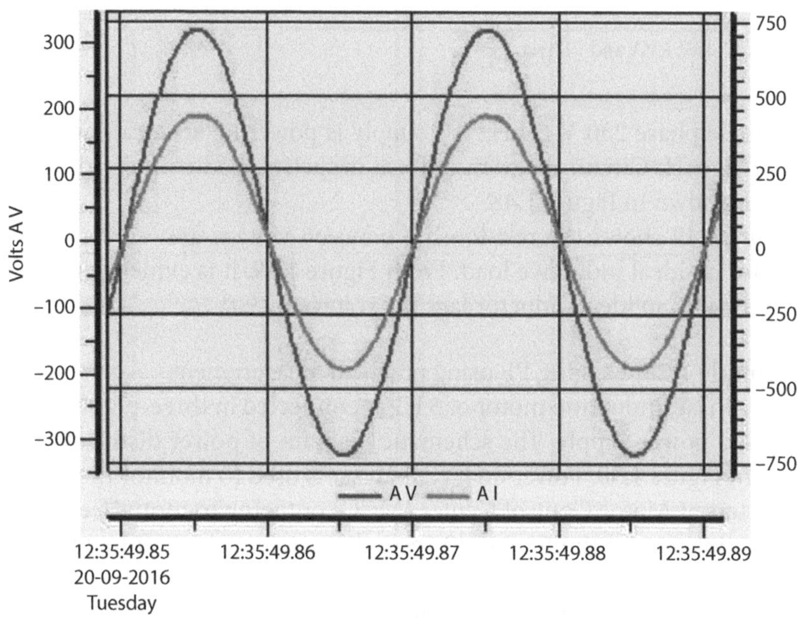


**Figure 1.46** Voltage and current wave shape for unity power factor load. Note: This figure is captured using Dranetz Power Quality analyser.

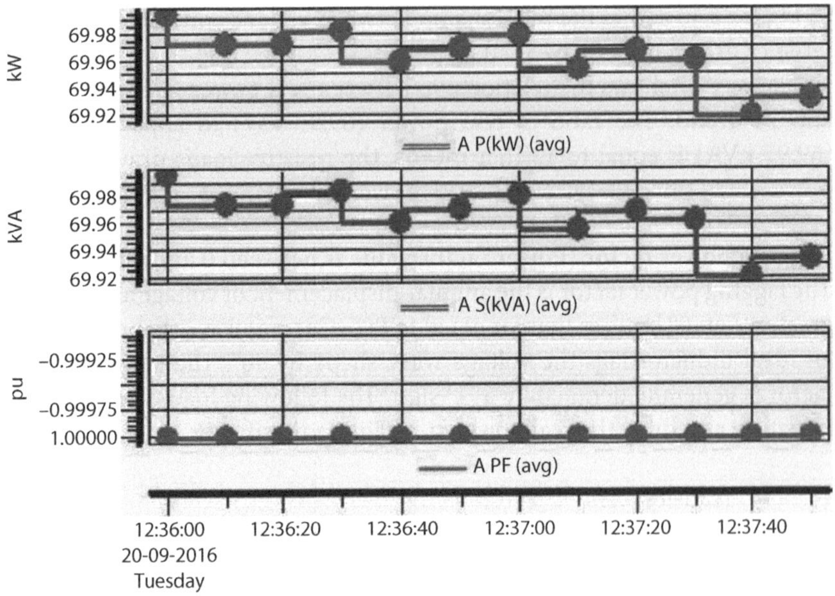


| | *Min* | *Max* | *Avg* |
|---|---|---|---|
| ***AP(kW)*** | *69.92* | *69.99* | *69.97* |
| ***AS(kVA)*** | *69.92* | *69.99* | *69.97* |
| ***APF*** | *1.0000* | *1.0000* | *1.0000* |

**Figure 1.47** kW, kVA and PF trend.

A single phase 230 V, 50 Hz AC supply is powering an ideal inductive load of 10 mH. Circuit diagram of ideal inductive load connected to AC supply is shown in Figure 1.48.

Figure 1.49 shows the relationship between voltage and current wave shape for an ideal inductive load. From Figure 1.49, it is evident that current drawn by an ideal inductor lags the voltage by 90°.

**Example 1.28:** Lagging PF using practical measurement
A three-phase, induction motor of 5 HP is connected in three-phase, 415V, 50 Hz AC power supply. The schematic diagram of power distribution is shown in Figure 1.50. Power quality analyzer is used to monitor the power parameters at Motor Control Center (MCC) outgoing to motor feeder for the duration is 11 minute and 40 seconds between 11:12:30 to 11:24:10 hours.

Figure 1.51 shows the relationship between voltage and current characteristics of 5 HP induction motor. From Figure 1.51, the current drawn by 5 HP induction is lags the voltage wave shape by 35.156°. This load is

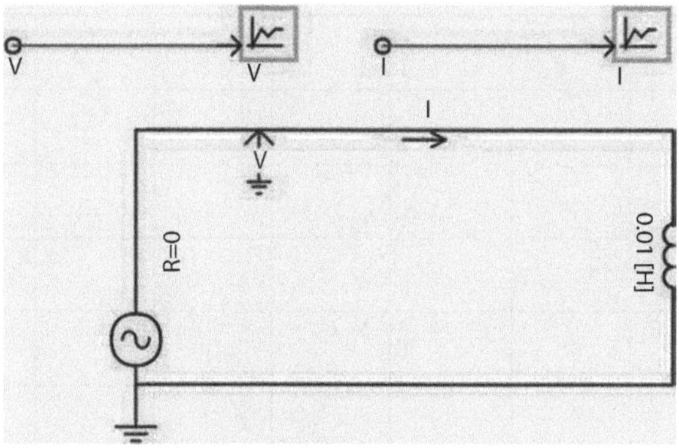


**Figure 1.48** Circuit diagram of ideal inductor.

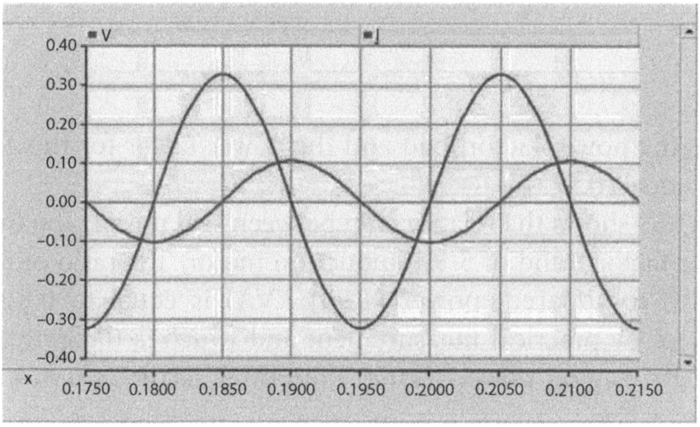


**Figure 1.49** Voltage and current relation for lagging power factor load.

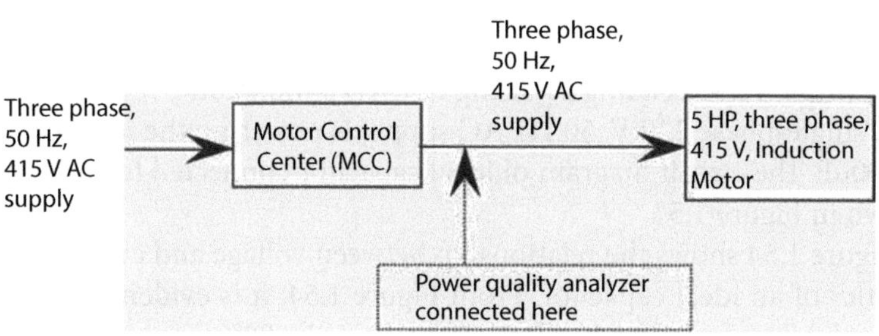


**Figure 1.50** Schematic diagram.

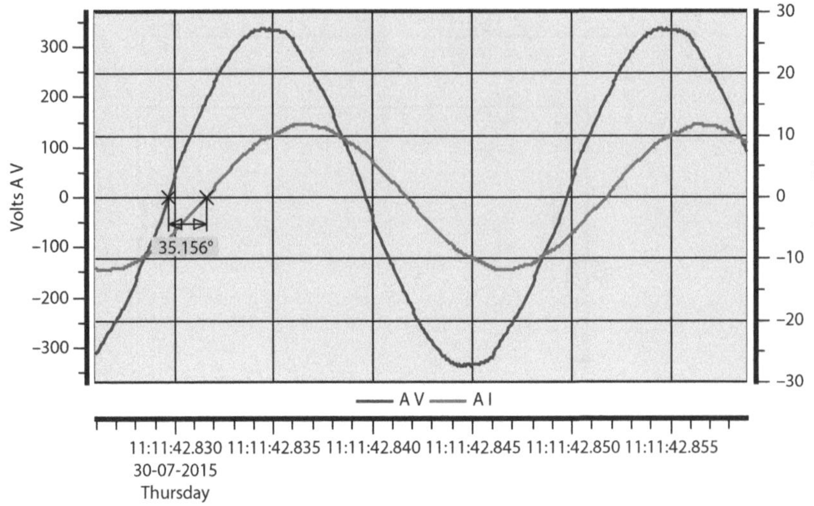


**Figure 1.51** Instantaneous voltage and current wave shape (R phase) of 5 HP induction motor.

called lagging power factor load and the power factor for this load is cos (35.156°) equals 0.82.

Figure 1.52 shows the relationship between real power, apparent power and power factor trend of 5 HP induction motor. The ratio of real power (1.597 kW) to apparent power (1.961 kVA) is equal to 0.814 lagging power factor by practical measurement and which is theoretically calculated as 0.82 lagging power factor. The power factor of 0.814 lagging is denoted as '+'0.81.

**Leading PF (PF value between -0.99 and 0):**

The leading PF is the angular displacement of voltage and current greater than 0° but less than or equal to 90°. The current drawn by an ideal capacitor leads the voltage by 90°. The leading PF is denoted by a '-' sign.

**Example 1.29:** Leading PF using PSCAD simulation

A single-phase 230 V, 50 Hz AC supply is powering the ideal capacitor of 500μF. The circuit diagram of ideal capacitor connected to AC supply is shown in Figure 1.53.

Figure 1.54 shows the relationship between voltage and current characteristics of an ideal capacitor. From Figure 1.54, it is evident that current drawn by the ideal capacitor leads the voltage by 90°.

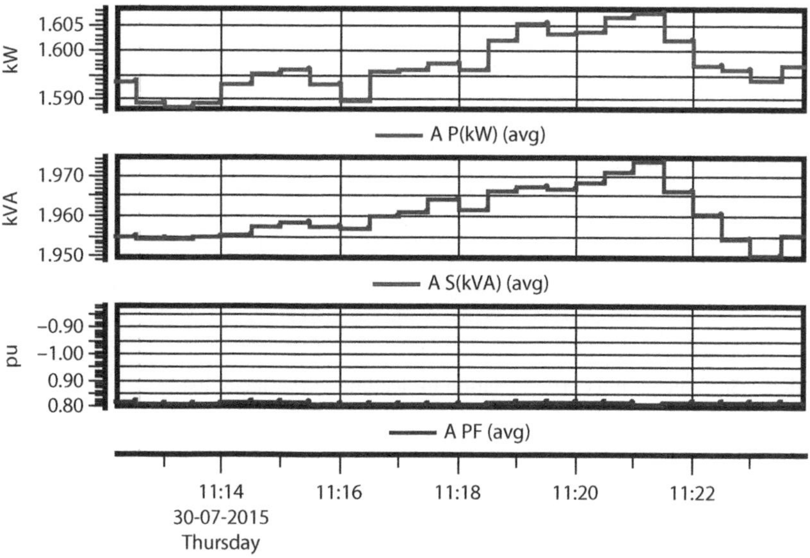


| | *Min* | *Max* | *Avg* |
|---|---|---|---|
| *AP(kW)* | *1.589* | *1.607* | *1.597* |
| *AS(kVA)* | *1.950* | *1.974* | *1.961* |
| *APF* | *0.8124* | *0.8174* | *0.8145* |

**Figure 1.52** Real, apparent power and power factor trend of 5 HP induction motor.

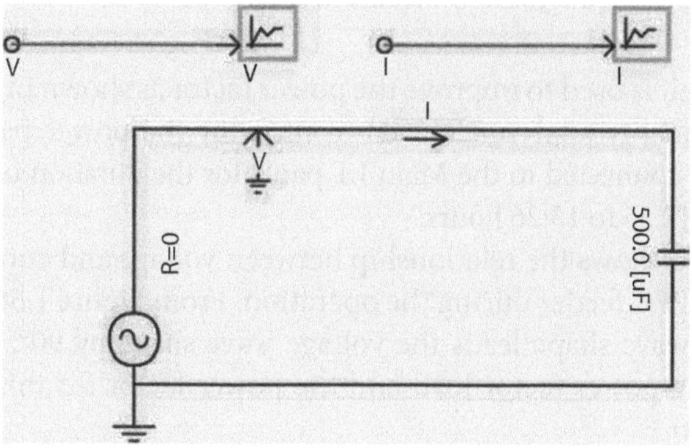


**Figure 1.53** Circuit diagram for an ideal capacitor.

**Example 1.30:** Leading power factor using practical measurement

A three-phase, 415 V, 50 Hz AC supply is powered to a combination of resistive and inductive loads (RL load). Automatic Power Factor Corrector

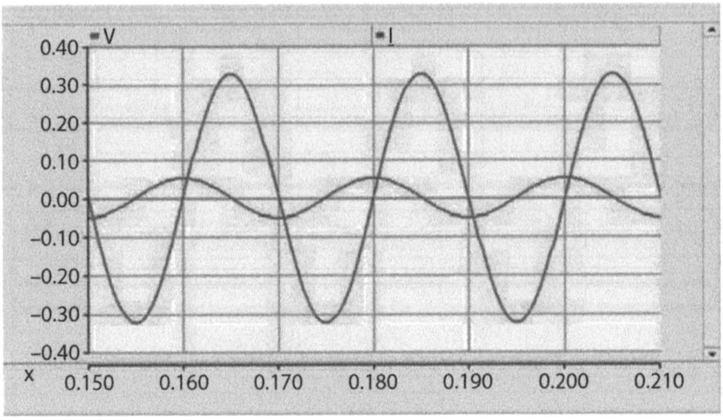


**Figure 1.54** Voltage and current relation for leading power factor load.

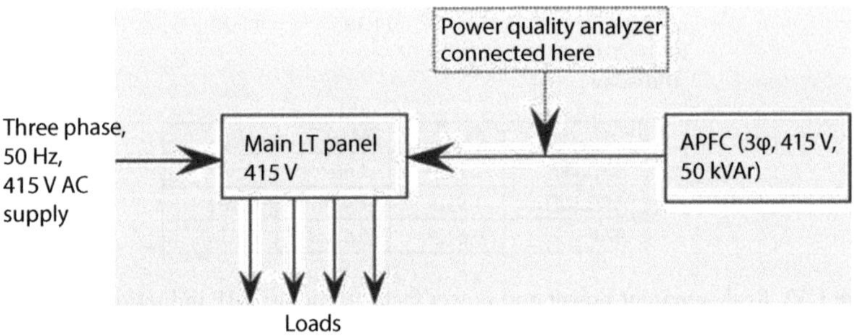


**Figure 1.55** Schematic diagram of power distribution.

(APFC) circuit is used to improve the power factor as shown in Figure 1.55. The power quality analyzer is used to monitor the power parameters of APFC feeder connected to the Main LT panel for the duration of 30 minutes data logging 12:56 to 13:26 hours.

Figure 1.56 shows the relationship between voltage and current characteristics of APFC feeder during the operation. From Figure 1.56, the capacitor current wave shape leads the voltage wave shape by 90°. This load is called leading power factor load and the power factor for this load is cos (90°) equals 0.

Figure 1.57 shows the relationship between active power and apparent power. The ratio of real power (0.17 kW) and apparent power (14.97 kVA) is equal to 0.01 leading power factor and which is theoretically calculated as 0. The power factor of 0.01 leading is denoted as '-'0.01.

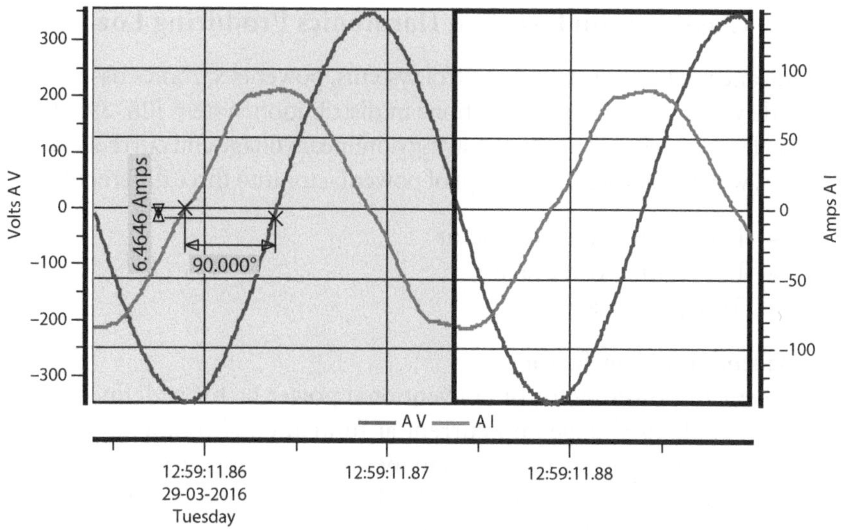


**Figure 1.56** Instantaneous voltage and current wave form of capacitor.

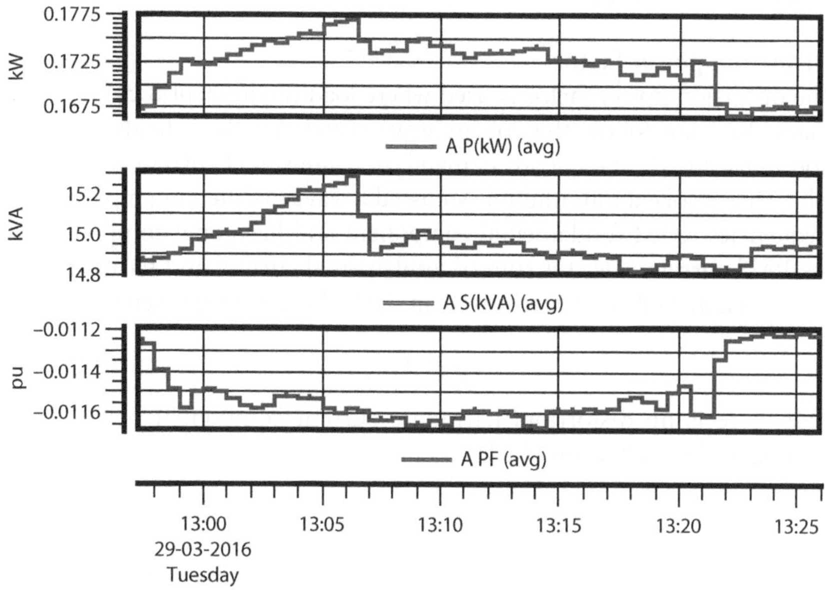


| | *Min* | *Max* | *Avg* |
|---|---|---|---|
| ***AP(kW)*** | *0.1668* | *0.1770* | *0.1723* |
| ***AS(kVA)*** | *14.82* | *15.29* | *14.97* |
| ***APF*** | *−0.01166* | *−0.01121* | *−0.01151* |

**Figure 1.57** kW, kVA and PF trend for leading PF load.

### 1.4.2 Classification Based on Harmonics Producing Loads

The advancement in electronic control systems, power electronics based loads are widely used in various applications in distribution system [28, 29]. These power electronics based loads are having nonlinear voltage and current characteristics, which changed the concept of power factor into three different forms:

- Displacement power factor
- Distortion power factor
- True power factor

**Displacement power factor:**
Displacement power factor or conventional power factor is defined as the cosine of angle of voltage and current at fundamental frequency without presence of harmonics. The displacement power factor is applicable for linear loads at fundamental frequency (50 Hz or 60 Hz only). If harmonics are present in the system, linear power factor is not the correct terminology to measure the power factor and distortion power factor that are to be used in such an environment.

**Distortion power factor:**
The distortion power factor is used where load (equipment's) current wave shape does not follow the voltage wave shape and these loads are called non-linear loads. These loads demand, non sinusoidal current wave shape, which are resolved into multiple sinusoidal wave of different amplitude and frequencies called fundamental component and harmonic component. As the RMS current is vector sum of fundamental current and harmonic current, fundamental current are less than the RMS current as given in eqn 1.23.

$$I_F < I_{RMS} \tag{1.23}$$

If there is no distortion, harmonic current component become zero leading to the expression (1.24)

$$I_F = I_{RMS} \tag{1.24}$$

Distortion power factor is the ratio upon fundamental current to RMS current. The more harmonic distortion, the less will be the distortion power factor. If the connected load is linear load, distortion power factor is unity. The expression for distortion power factor is given in eqn 1.25.

$$\text{PF (distortion)} = \frac{I_F}{I_{RMS}} \tag{1.25}$$

Where
$I_F$ is fundamental current
$I_{RMS}$ is total current

**True power factor:**
True power factor is the overall power factor of the presence of both linear and non-linear (harmonics) loads. It is the product of the displacement power factor and distortion power factor. The expression for true power factor is given in eqn. 1.26.

$$\text{PF (true)} = \text{PF (displacement) X PF (distortion)} \tag{1.26}$$

When the connected loads are linear, both displacement and distortion power factor is unity and true power factor is unity.

### 1.4.3 The Need for Power Factor Improvement

The drawl of entire reactive power demanded by the plant loads from the power source (DISCOMs) affects the power factor limits specified by the particular DISCOMs. Drawl of real power at low power factor will increase the apparent power requirement of the plant and additional demand capacity to be purchased from the DISCOMs. This could be reduced by supplying the localised reactive power to the loads as a localised compensation and increase the real power consumption at the same apparent power from the DISCOMs power supply. Supplying the localised reactive power to increase real power has the following advantages in the system:

- Reduces the apparent power demand of the system
- Avoids the penalty imposed by the power supply company or DISCOMs
- Reduces electricity cost
- Reduction of circuit components rating like cable, switchgear, etc
- Reduction of $I^2R$ losses in the system equipment's like cables
- Reduces the voltage drop in the cable

### 1.4.4 Methods of Power Factor Improvement

The power factor improvement are obtained majorly by two ways:

1. Capacitor banks
2. Synchronous condenser

**Power factor improvement by installing the capacitor banks:**
The capacitor banks are an economical way to produce the reactive power requirements. These capacitor banks are used to improve the power factor by supplying the required reactive power locally in the power system. It Improves power factor by reducing the phase angle between voltage and current by supplying the reactive power demanded by inductive loads [19]. Generally, these capacitor banks are connected parallel to the loads which require the reactive power for their operation.

The capacitor banks are used to improve the power factor as shown in Figure 1.58.

In Figure 1.58, there are multiple Motor Control Centers (MCC) that are connected in MV switchgear at 11 kV voltage level. This MCC has multiple motors which operate in different load cycles in 24 hours. These motors require reactive power for their operation. The Automatic Power Factor Capacitor (APFC), is connected at 11 kV MV switchgear which is used to supply the varying reactive power required by MCC-1 to MCC-5.

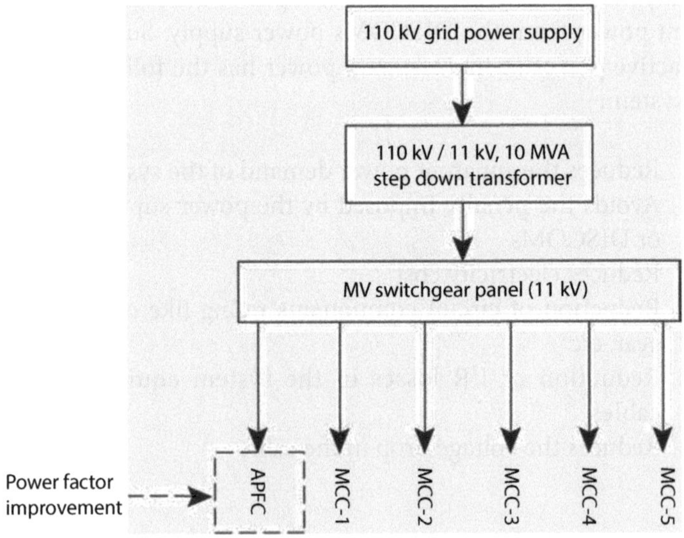


**Figure 1.58** Power factor improvement by capacitor bank.

**Synchronous condenser:**

The synchronous condenser is one of the methods for power factor compensation in the transmission and distribution system. The synchronous motor has the capability of operating in all the power factor (lagging, leading or unity) based on the excitation. Whenever the synchronous motor is operating at no load and over excited condition, this draws the real power from the power supply and it injects the reactive power to the power supply to improve the power factor. Nowadays, power factor compensation by synchronous condenser methods are not widely used because of higher costs involved.

The synchronous condenser is used to improve the power factor; this is represented in Figure 1.59.

## 1.5 Types of Loads

The electrical loads are classified into two types based on the voltage and current relationship:

- Linear loads
- Non-linear loads

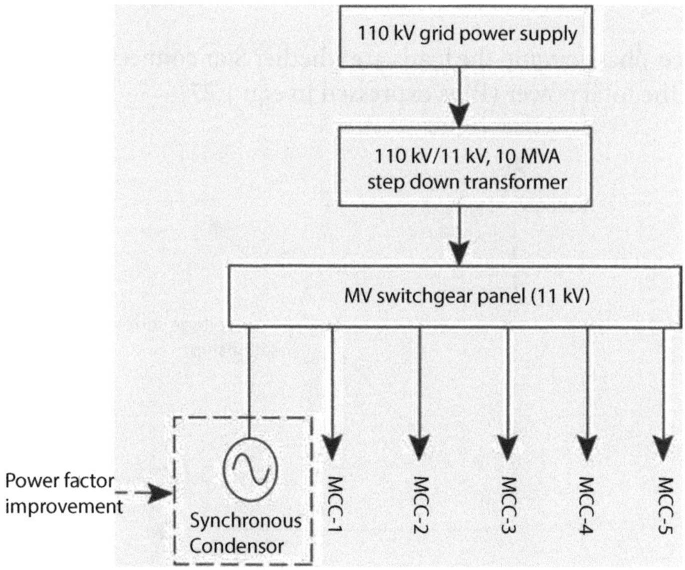


**Figure 1.59** Power factor improvement by synchronous condensor.

### 1.5.1 Linear Loads

Linear loads are loads which draw a sinusoidal current pattern when sinusoidal voltage is applied. Linear loads operate at fundamental frequency of 50 Hz or 60 Hz only. In other words V/I is a straight line which means the loads that obey Ohm's law is a linear load. The Figure 1.60 shows the voltage and current relationship of linear loads.

Examples for linear loads are electric heater, incandescent lamp and electric motor.

### 1.5.2 Non-Linear Loads

Non-linear loads are loads which draw a non-sinusoidal current pattern when sinusoidal voltage is applied across its terminals. The characteristics of non-linear loads are explained in section 1.1.4.2 under distortion power factor. Figure 1.61 shows the voltage and current relationship of non-linear loads.

Example for non-linear loads are power electronic based equipment's like VFD, UPS, SMPS and Cyclo converter, etc.

The voltage - current relationship for VFD, UPS and SMPS are shown in Table 1.2.

## 1.6 Three-Phase Power Measurement

In a three-phase circuit, the loads are whether Star connected or Delta connected, the total power (P) is expressed in eqn 1.27.

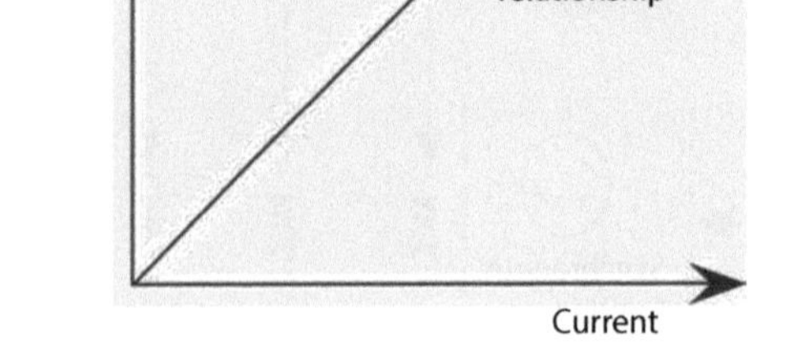


**Figure 1.60** Linear voltage – current relationship.

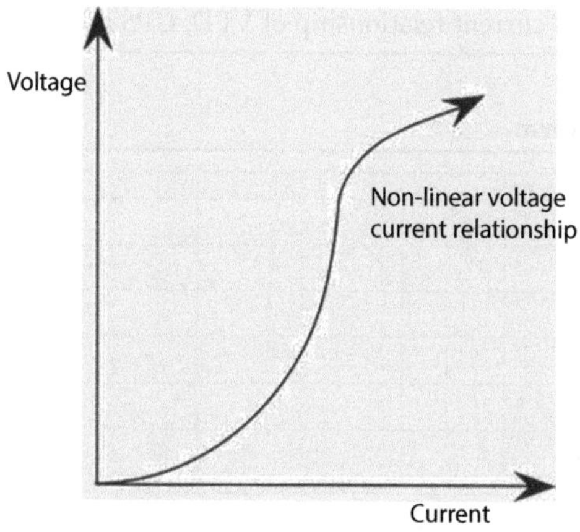


**Figure 1.61** Non-linear voltage – current relationship.

$$P = \sqrt{3} V_L I_L \cos(\phi) \tag{1.27}$$

Where

P is power in Watts

$V_L$ is line voltage

$I_L$ is line current

cos ($\phi$) is power factor

The Wattmeter is a measurement device, which is used to measure the power in a single-phase and/or three-phase system. It consists of two coils, namely current coil and voltage coil.

(i) Current Coil

The current coil is connected in series with the load and used to sense the current flowing in the circuit. It is similar to ammeter and the resistance of the coil is kept as low as possible. Hence, it requires higher cross sectional area and less number of turns. The current coil is shown in Figure 1.62.

(ii) Voltage Coil

The voltage coil or pressure coil is connected in parallel with the load and used to sense the voltage across the power supply terminals. It is similar to voltmeter and the resistance of the coil is as high as possible. Hence, it requires less cross sectional area and a greater number of turns. The pressure coil is shown in Figure 1.63.

**Table 1.2** Voltage – current relationship of VFD, UPS and SMPS.

| Type of load | Waveform |
|---|---|
| VFD | 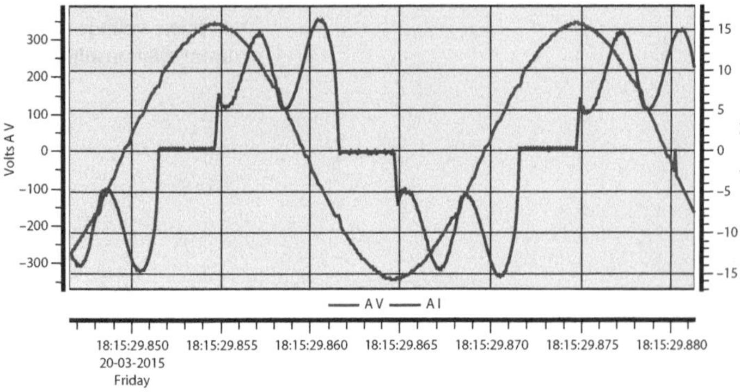 |
| UPS | 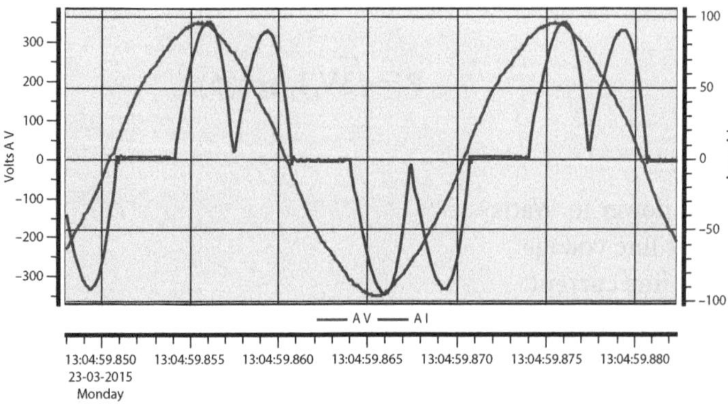 |
| SMPS | 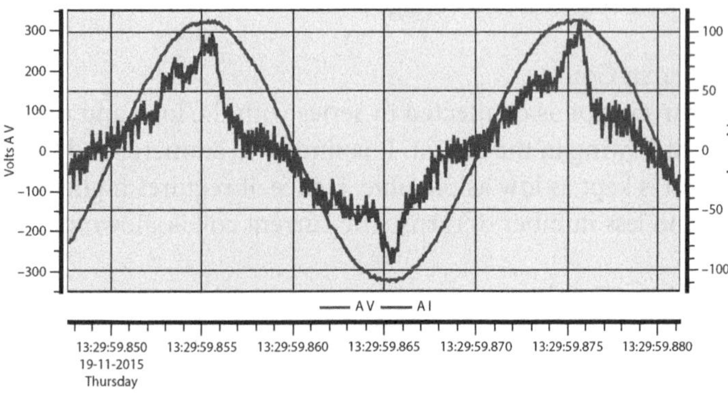 |

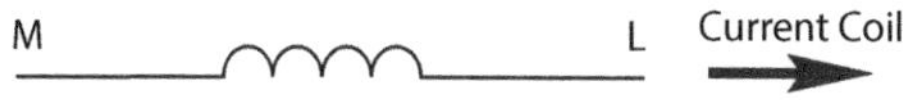


**Figure 1.62** Current coil.

**Figure 1.63** Pressure coil.

Where
M is from mains
L is to load (for current coil)
C is common
V is voltage (pressure coil)

In general, power in three phase circuit is measured by the following methods:

1) Two-Wattmeter method
2) Three-Wattmeter method

**A. Two-Wattmeter Method:**
The two-Wattmeter method is generally used to measure the power in a three-phase, three-wire circuit; it is shown in Figure 1.64. The two Wattmeter are namely $W_1$ and $W_2$ connected to either two phase or three phase.

The Wattmeter $W_1$ is connected in R and B phase and Wattmeter $W_2$ is connected in Y and B phases. In Wattmeter $W_1$, current coil terminal M is connected to R phase and L is connected to load terminal of R phase. Voltage coil terminal C is connected to M and V is connected to B phase.

In Wattmeter $W_2$, current coil terminal M is connected to Y phase and L is connected to load terminal of Y phase. Voltage coil terminal C is connected to M and V is connected to B phase.

Similar type of connection is used for delta connected load.

In this method, total power in Watts is the algebraic sum of Wattmeter $W_1$ and Wattmeter $W_2$ expressed in equation 1.28.

$$W = W_1 + W_2 \tag{1.28}$$

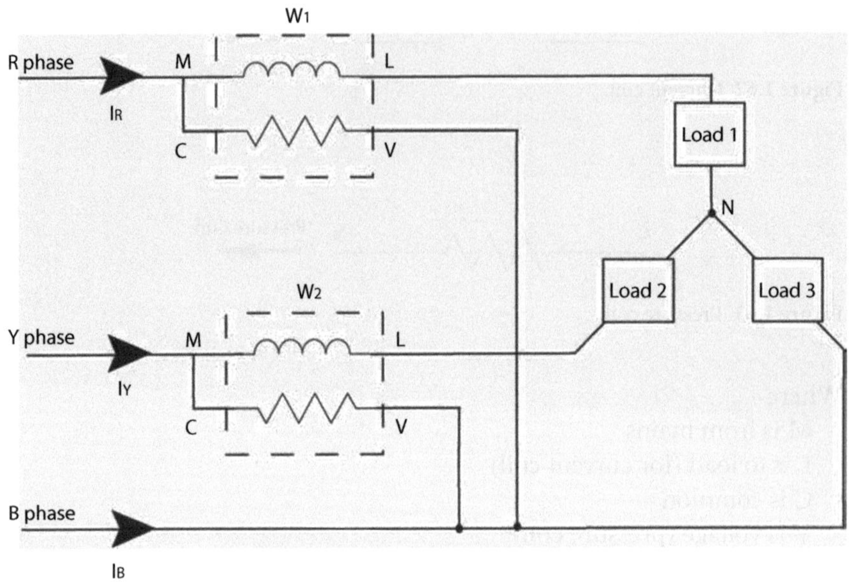


**Figure 1.64** Two-Wattmeter method for three-phase power measurement.

Where

W is total power in (W)

$W_1$ is power (W) in Wattmeter 1

$W_2$ is power (W) in Wattmeter 2

The advantages of the two-wattmeter method are as follows

- The method is suitable for both balanced and unbalanced loads
- Neutral point for star connected load is not necessary to connect the Watt meters
- Two-Wattmeter's are sufficient to measure the total power

The disadvantages of two-Wattmeter methods are as follows

- This method is not suitable for three-phase, four-wire loads
- Higher possibilities of measurement error due to sign of both Wattmeter $W_1$ and $W_2$

**B. Three-Wattmeter method:**

The three-Wattmeter methods are generally used to measure the power in three-phase, three-wire circuits and three-phase, four-wire circuits.

The connections for star connected loads for measuring power by the three-wattmeter method is shown in the following Figure 1.65. The three-wattmeters are namely $W_1$, $W_2$ and $W_3$.

The Wattmeter $W_1$ is connected in R phase and neutral, Wattmeter $W_2$ is connected in Y phase and neutral, Wattmeter $W_3$ is connected in B phase and neutral. In Wattmeter $W_1$, current coil terminal M is connected to R phase and L is connected to load terminal of R phase. Voltage coil terminal C is connected to M and V is connected to neutral.

In Wattmeter $W_2$, current coil terminal M is connected to Y phase and L is connected to load terminal of Y phase. Voltage coil terminal C is connected to M and V is connected to neutral.

In Wattmeter $W_3$, current coil terminal M is connected to B phase and L is connected to load terminal of B phase. Voltage coil terminal C is connected to M and V is connected to neutral.

In this method, total power in Watts is the algebraic sum of wattmeter $W_1$, wattmeter $W_2$, wattmeter $W_3$ and expressed in equation 1.29.

$$W = W_1 + W_2 + W_3 \tag{1.29}$$

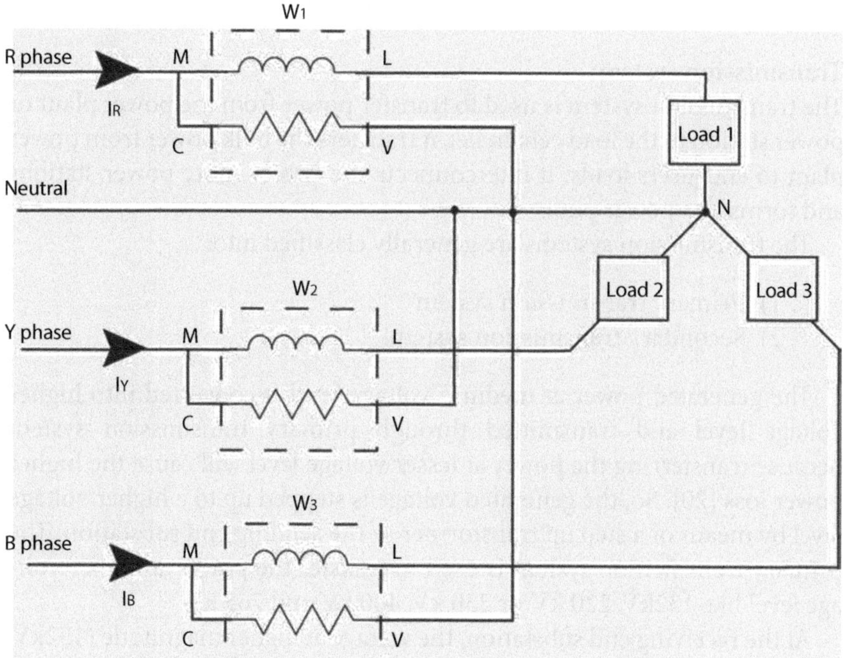


**Figure 1.65** Three-Wattmeter method power measurement.

Where

W is total power in (W)
$W_1$ is power (W) in wattmeter 1
$W_2$ is power (W) in wattmeter 2
$W_3$ is power (W) in wattmeter 3

## 1.7 Overview of Power Systems

The power plant or power station, power transmission and power distribution systems are the main parts of an electrical power system [2]. The generated power at a power plant, in a nuclear, coal or solar power plant, is transmitted to the loads through transmission and distribution systems [16, 21]. The general structure of the power system is shown in Figure 1.66.

**Power plant or power station:**
The power plant generates the power from conventional sources like coal, nuclear, diesel, etc., and non-conventional sources like solar, wind, biomass, etc [15, 32–34]. The voltage level of the power generation at the power plant is between 6.6 kV to 22 kV based on the requirement.

**Transmission system:**
The transmission system is used to transfer power from the power plant or power station to the load center, i.e., it transfers the bulk power from power plant to end-users loads. It interconnects the two or more power stations and forms as a power pool.

The transmission systems are generally classified into:

1) Primary transmission system
2) Secondary transmission system.

The generated power at medium voltage level is converted into higher voltage level and transmitted through primary transmission system because transferring the power at lesser voltage level will cause the higher power loss [20]. So, the generated voltage is stepped up to a higher voltage level by means of a step up transformer in the sending end substation. The primary transmission system is used to transfer the power at higher voltage level like 132kV, 220 kV or 230 kV, 400 kV and 765 kV.

At the receiving end substation, the voltage at higher magnitude (132kV, 220 kV or 230 kV, 400 kV and 765 kV) is stepped down to 66 kV or 33 kV by means of a step down transformer. The secondary transmission system

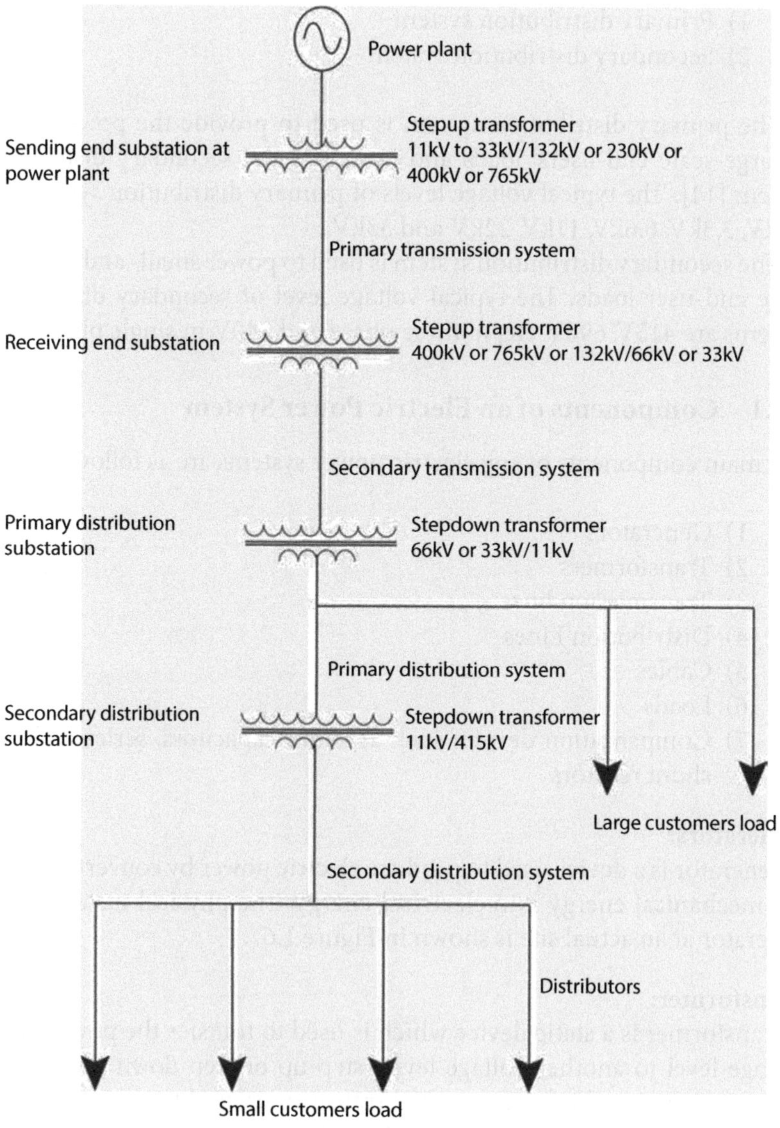


**Figure 1.66** General structure of power system.

is used to transfer the power to the primary distribution system and large-scale end user loads at 33 kV or 66 kV.

**Distribution system:**
The distribution system is used to provide the power supply to end-users' loads [9, 13, 18]. It generally classified into two types:

1) Primary distribution system
2) Secondary distribution system

The primary distribution system is used to provide the power supply to large-scale end-users' loads and powering the secondary distribution system [14]. The typical voltage levels of primary distribution systems are 2.2kV, 3.3kV, 6.6kV, 11kV, 22kV and 33kV.

The secondary distribution system is used to power small- and medium-scale end-user loads. The typical voltage level of secondary distribution systems are 415V, 690V, etc, in three phase and 240V in single phase.

### 1.7.1 Components of an Electric Power System

The main components of any electric power systems are as follows [4, 5]:

1) Generators
2) Transformers
3) Transmission lines
4) Distribution Lines
5) Cables
6) Loads
7) Compensation devices such as shunt capacitors, series and shunt reactors

**Generators:**
A generator is a device used to produce electric power by converting rotating mechanical energy into electrical energy. The physical installation of generator at an actual site is shown in Figure 1.67.

**Transformer:**
A transformer is a static device which is used to transfer the power at one voltage level to another voltage level (step up or step down the voltage) without changing the frequency. The transformer used to change the voltage from lower voltage to higher voltage is called a step-up transformer, and the transformer that is used to reduce the voltage level from higher voltage level to lower voltage level is called a step-down transformer.

The physical installation of a transformer at an actual site is shown in Figure 1.68.

**Transmission Lines:**
Overhead transmission lines are used to transfer electric power from one location to another at higher voltage level in a primary transmission

**Figure 1.67** Installation of generator at actual site.

**Figure 1.68** Installation of transformer at actual site.

system and a secondary transmission system. Based on number of circuits, the transmission lines are classified as follows:

i) Single-circuit transmission line
ii) Double-circuit transmission line
iii) Four-circuit transmission line

The single-circuit transmission line consists of one circuit of R, Y and B phases. The actual site installation of single-circuit transmission lines is shown in Figure 1.69.

The double-circuit transmission line consists of two numbers of single circuit of R, Y and B phases. Actual site installation of double-circuit transmission lines is shown in Figure 1.70.

**Figure 1.69** Installation of transmission lines 110 kV single circuit.

**Figure 1.70** Installation of transmission lines 220 kV double circuit.

The four-circuit transmission line consists of four numbers of single circuit of R, Y and B phases. Actual site installation of four-circuit transmission lines is shown in Figure 1.71.

The distribution lines are used to transfer electric power from one location to another location at medium voltage level in a primary distribution system and low voltage level in a secondary distribution system [12]. The actual site installation of a medium voltage primary distribution line is shown in Figure 1.72.

The actual site installation of low voltage secondary distribution line is shown in Figure 1.73.

**Figure 1.71** Installation of transmission lines.

## Cables:

Power cables are a replacement for transmission lines in various places to transfer the power from one location to another location. These power cables are used to transfer electric power from one location to another location in both transmission and distribution system. Generally, power cables are installed in cable trays or conduits or in trenches. The cables are buried; these cables are called underground cables.

The actual site installation of underground cable for medium voltage primary distribution system is shown in Figure 1.74.

**Figure 1.72** Installation of primary distribution line 11 kV single circuit.

The selection installation whether overhead lines or underground cables are depends on the following factors:

1) Location of installation
2) Technical and commercial considerations

The comparison of both overhead transmission system and underground transmission system are listed and given in Table 1.3.

## 1.8 Protection of Power System

The power system protection deals with the safety of the electrical equipment's in the power system from faults by isolating the faulty portion from other healthy equipment's in the system [17]. When abnormalities (faults) are present in the system, current flow in the (line or cable) circuit increases drastically. Compared to normal operation current, current flow during

**Figure 1.73** Installation of secondary distribution line 415 V.

the faulted condition is very high (more than several times of full load current). The equipment's get damaged due to this high current flow during the faulted condition. Hence, it is essential to protect the equipment's from the high current during the fault condition. The aim of the power system protection is to maintain the equipment's safety and stability of the power system by isolating the faulty equipment's during the system faults, ensuring the continuous operation of healthy equipment's in the network. The devices which are used to protect the system from faults are called switchgears or protection devices.

The combination of Current Transformers (CTs), Potential Transformers (PTs), relays and circuit breakers are used to achieve the protection. The power system protection is selected for low voltage, medium voltage and high voltage based on the system operating voltage [23]. The general flow chart of power system protection is shown in Figure 1.75.

The basic components of power system protection are listed as follows:

**Figure 1.74** Installation of underground cables in buried cable trench.

1) Current Transformers (CTs) and Voltage Transformers (VTs)
2) Protective Relays
3) Circuit Breakers
4) Other accessories like batteries, charger and communication interface, etc.

**Current Transformers:**

Current transformers are used to reduce the high current magnitude flowing in the circuit to lesser current magnitude to a convenient level suitable

**Table 1.3** Comparison of overhead and underground transmission system.

| S. No. | Overhead system | Underground system |
|---|---|---|
| 1 | De-rating factor is less | De-rating factor is high |
| 2 | The required amount of insulation for same voltage level is less | The required amount of insulation for same voltage level is high |
| 3 | Heat dissipation of the conductor is easy | Heat dissipation of the conductor is difficult |
| 4 | Lesser economics involved in same power rating | Higher economics involved in same power rating |
| 5 | Fault identification is easy | Fault identification is difficult |
| 6 | Maintenance is simple | Maintenance is complicated |
| 7 | Long distance transmission | Short distance transmission |
| 8 | Safety is less | Safety is high |
| 9 | Affected by lightning discharge | Not affected by lightning discharge |
| 10 | Highly affected by natural disaster | Less affected by natural disaster |

for the operation of metering instruments and/or relays. It has two windings, namely primary winding and secondary winding. Primary winding is connected at high current side and secondary winding is connected to metering equipment's and relays. The installation of CTs at 11 kV medium voltage is shown in Figure 1.76.

The primary current of the CTs is based on the current flow in the particular circuit, and secondary current rating of the CTs are either 1 A or 5 A. For example, the current flow in the circuit is 100 A and required secondary current is 1 A, then the required CT ratio is 100/1 A. Figure 1.77 shows the typical 100/1 A two core CT, core 1 is accuracy class 1 for metering and core 2 is accuracy class 5P20 for protection.

The construction requirements for both metering and protection requirements are different. Thus, metering CTs are not suitable for protection purpose and protection CTs are not suitable for metering purpose.

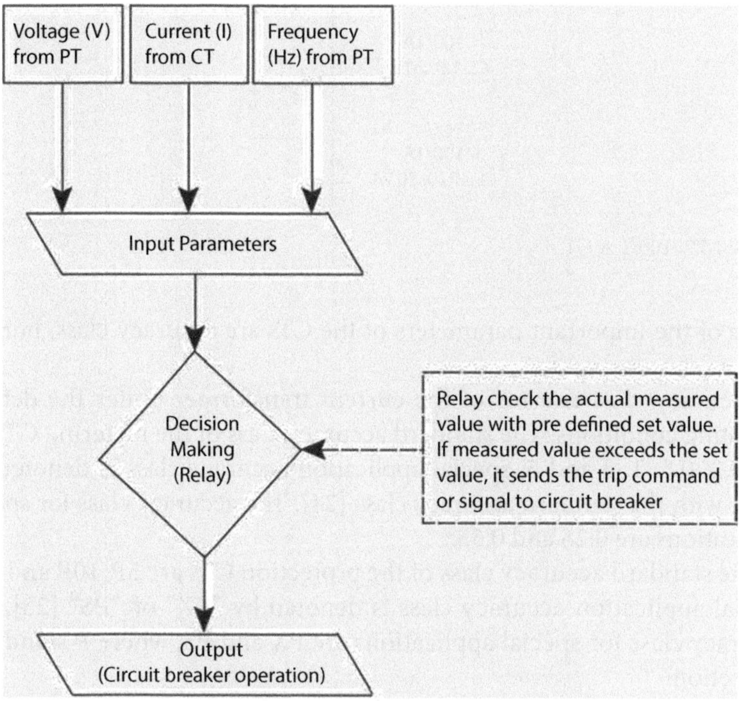


**Figure 1.75** Flow chart of power system protection.

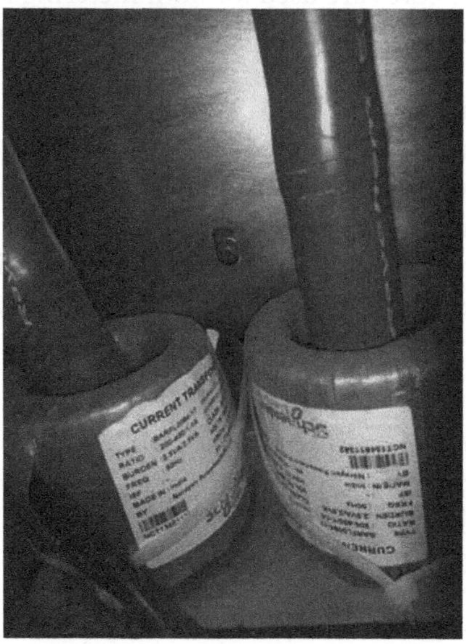

**Figure 1.76** CTs at 11 kV (Courtesy: Schneider Electric).

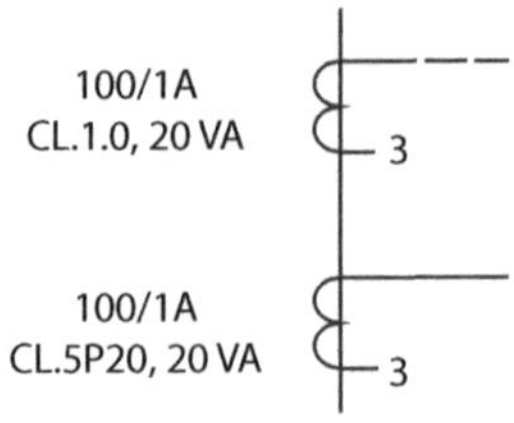


**Figure 1.77** 100/1 A CT.

Some of the important parameters of the CTs are accuracy class, burden, etc.

Accuracy class is error of the current transformer under the defined operating conditions. The standard accuracy class of the metering CTs are 0.1, 0.2, 0.5, 1, 3 and 5, special application accuracy class is denoted "S" along with the standard accuracy class [24]. The accuracy class for special applications are 0.2S and 0.5S.

The standard accuracy class of the protection CTs are 5P, 10P and 15P, special application accuracy class is denoted by "PX" or "PS" [25]. The accuracy class for special applications are PX and PS, where P stands for protection.

Burden is secondary circuit impedance in Ohms and power factor of the connected equipment's like relays, meters, transducers, etc.

Here the multiple cores are used for both metering and protection purposes; name plate details of the multi-core CT are listed in Table 1.4.

**Voltage Transformers:**

Voltage transformers are used to reduce the high voltage magnitude in the circuit to lesser voltage magnitude to a convenient level suitable for operation of metering instruments and/or relays. Voltage transformers are also called Potential Transformers (PTs). It has two windings, namely primary winding and secondary winding. Primary winding is connected at high voltage side

**Table 1.4** Name plate details of the multi-core CT.

| S. No. | Specification | Metering core | Protection core |
|---|---|---|---|
| 1 | Ratio (A) | 400/1 | 400/1 |
| 2 | Accuracy class | 0.5 | 5P10 |
| 3 | Burden (VA) | 5 | 5 |

**Figure 1.78** VTs at 11 kV.

and secondary winding is connected to metering equipment's and/or relays. The typical PT for 11 kV medium voltage is shown in Figure 1.78.

The primary voltage of the PTs is selected based on the voltage at the point of connection and secondary voltage rating is 110V.

Like CTs, the construction requirements for both metering and protection requirements are different. Thus, metering PTs are not suitable for protection purpose and protection PTs are not suitable for metering purpose. Some of the important parameters of the PTs are accuracy class, burden, etc.

Accuracy class is the error of the potential transformer under the defined operating conditions. The standard accuracy class of the metering PTs are 0.1, 0.2, 0.5, 1 and 3 [27]. The standard accuracy class of the protection PTs are 3P, 6P [26].

Burden is secondary circuit impedance in Ohms of the connected equipment's like relays, meters, transducers, etc.

The name plate details of the multi-core PT in which the multiple cores for both metering and protection purpose are listed in Table 1.5.

**Table 1.5** Name plate details of the multi-core PT.

| S. No. | Specification | Metering core | Protection core |
|---|---|---|---|
| 1 | Ratio (V) | 33000/110 | 33000/110 |
| 2 | Accuracy class | 0.5 | 3P |
| 3 | Burden (VA) | 15 | 15 |

**Protective Relays:**

Protection relays are used to sense the abnormalities (fault) from the inputs (voltage, current and frequency). If the relay senses the abnormalities, then it will initiate the trip command to circuit breaker to isolate the faulty portion from the system. There are different type of relays that are used in a power system to identify the abnormalities, such as

1) Current
2) Voltage
3) Frequency
4) Impedance
5) Direction of current flow

   Etc.

Based on the application, different types of relays are required to protect the equipment's during the system faults. They are listed below:

- Over current and earth fault relay
- Over voltage and under voltage relay
- Over frequency and under frequency relay
- Differential relay
- Distance relay
- Directional relay

  Etc.

**Figure 1.79** Instantaneous earth fault relay (Courtesy: Alstom).

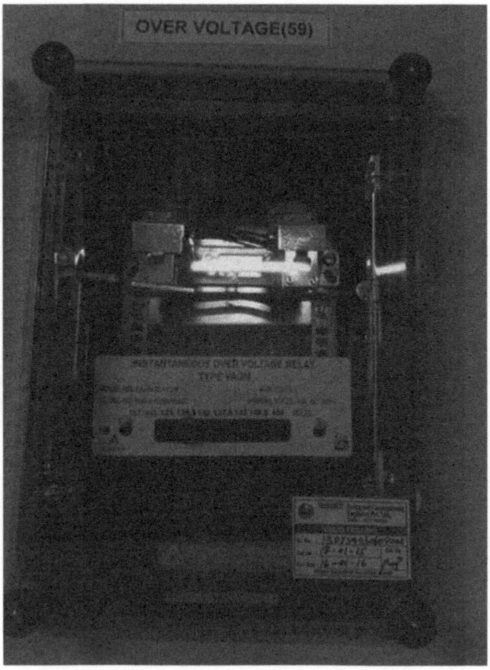


**Figure 1.80** Instantaneous over voltage relay (Courtesy: Areva).

The instantaneous earth fault relay is shown in Figure 1.79.

The instantaneous over voltage relay is shown in Figure 1.80.

**Circuit Breaker:**

A circuit breaker is a device which is used to open and close the circuit contacts by means of manual as well as automatic (electrical) operation. There are different types of circuit breakers that are used for protection purposes: Vacuum Circuit Breaker (VCB), Sulfur Hexa Fluoride ($SF_6$), etc., in a high voltage system.

The actual installation of an SF6 breaker at site is shown in Figure 1.81.

**Other accessories:**

Batteries are used to provide the continuous and uninterrupted auxiliary power supply to relays, metering instruments and circuit breakers. Communication channels are used to analyse the voltage and current at the remote end of a line and tripping the circuit using the circuit breaker in case of abnormalities.

In a modern power system, disturbance recorders are used to monitor the system disturbance during the faulted conditions.

**Figure 1.81** Installation of $SF_6$ breaker.

The power system protection has the following performance measures to check the effectiveness of the protection:

- Selectivity
- Speed
- Reliability
- Economy
- Simplicity

**Basic Concepts:**

**Current:** The movement of charges or flow of electrons in a circuit is called current. Unit of the current is "Ampere".

**Voltage:** The potential difference between the two terminals in the circuit is called voltage. Unit of the voltage is "Volts".

**Power:** The rate at which the work is being done in an electrical circuit is called an electric power. The power is measured in terms of "Watts".

**Energy:** The power and time required to complete the work is called energy. It is measured in terms of "Watt hour". Energy = Power*time (Whr)

**Single-Phase Power Supply:** A single-phase power supply has one line conductor and one neutral conductor. For example, the potential difference between the line conductor and neutral is 230 Volts.

**Three-Phase Power Supply:** A three-phase power supply has three line conductors and/or one neutral conductor. For example, the potential difference between the two line conductors is 415 V and potential difference between line conductor to neutral conductor is $1/\sqrt{3}$ times of line to line voltage which is 240 V. The phase difference between the three phases are 120 degrees electrical each other or by 1/3rd of the time period. That is in a 50 Hz supply has 6.6 milliseconds lag between each of the three phases crossing a certain voltage value.

**Star Connection:** The Star connection is formed by connecting starting or terminating ends of all the three windings together. The ends $R_1$-$Y_1$-$B_1$ are connected or ends $R_2$-$Y_2$-$B_2$ are connected together. This common point is called Neutral point. The remaining three ends are brought out for connection purpose. These ends are generally referred to as R-Y-B, to which loads are to be connected.

**Delta Connection:** The Delta is formed by connecting one end of the winding to starting of the other winding end connection which are made to form a closed loop. The supply terminals are taken out from the three junction points. Delta connection always forms a closed loop.

**Line Voltage:** The potential difference between the two phases.

**Phase Voltage:** The potential difference between the phase and neutral.

**Line Current:** The current passing through line (only line to line).

**Phase Current:** The current passing through the load (i.e., Phase to load).

**Balanced Load:** The impedance of all the three phases are equal. It is also called symmetrical load.

**Unbalanced Load:** The impedance of all the three phases are not equal. It is also called unsymmetrical load.

**Methods of Three-Phase Power Measurements:**

i. Single-watt meter method
ii. Two-watt meter method
iii. Three-watt meter method

**Main Components of the Electrical Power System:**

i. Generator
ii. Transformers
iii. Transmission line (primary transmission and secondary transmission)
iv. Distribution line (Primary distribution and secondary distribution)
v. Control equipment's

**Types of Electrical Transmission Systems:**

i. Overhead transmission systems
ii. Underground transmission systems

**Protection of Power System:** The power system protection is a branch of electrical power engineering that deals with the protection of electrical power systems from faults by isolating the faulty parts from the rest of the electrical network.

**Power Factor:** The power factor is defined as the ratio of real power to the apparent power.

$$\text{Power factor} = \text{Real power}/\text{Apparent power}$$

**Why the power factor has to be maintained at around unity level:** In general, most of the loads we use are inductive loads, so the power factor is always less than unity power factor. If the power factor is reduced, then the availability of real power in the same apparent power is reduced. Hence, we need to maintain the power factor at around the unity.

**Benefits of Power Factor Improvement:**

- Reduces the apparent power demand of the system
- Avoids the penalty imposed by the power supply company or DISCOMs
- Reduces electricity cost
- Reduction of circuit components rating like cable, switch-gear, etc.
- Reduction of $I^2R$ losses in the system equipment's like cables
- Reduces the voltage drop in the cables

**Methods of Power Factor Improvement:**

- Capacitors banks
- Synchronous condenser

## References

1. Richard C. Dorf and James A. Svoboda, *Introduction to Electric Circuits*, 9th Edison, Wiley, 2018.
2. D. P. Kothari and I J Nagrath, *Modern Power System Analysis*, 4th Edison, Tata McGraw Hill, New Delhi, 2011.
3. D. P. Kothari, K. C. Singal and Rakesh Ranjan, *Renewable Energy Sources and Emerging Technologies*, Second Edison, PHI, 2011.
4. IEEE Std 141-1993 IEEE recommended practice for electric power distribution for industrial plant.
5. IEEE Std 241-1990 IEEE recommended practice for electric power systems in commercial buildings.
6. D. P. Kothari and I. J. Nagrath, *Electric Machines*, 4th Edison, Tata McGraw Hill, New Delhi, 2010.
7. P. Sivaraman, C. Sharmeela and D. P. Kothari, "Enhancing the voltage profile in distribution system with 40 GW of solar PV rooftop in Indian grid by 2022: a review" 1st International conference on Large scale grid integration renewable energy in India, September, 2017, New Delhi, India.
8. Power factor correction and harmonic filtering in electrical plants, Technical application papers, ABB, July 2018.

9. Juan M Gers, *Distribution system analysis and automation*, IET, 2013.
10. Sivaraman, P., and Sharmeela, C. (2020). Solar Micro-Inverter. In J. Zbitou, C. Pruncu, & A. Errkik (Eds.), *Handbook of Research on Recent Developments in Electrical and Mechanical Engineering* (pp. 283–303). Hershey, PA: IGI Global.
11. Sivaraman, P., and Sharmeela, C. (2020). Introduction to electric distribution system. In Baseem Kahn, Hassan Haes Alhelou & Ghassan (Eds.), *Handbook of Research on New Solutions and Technologies in Electrical Distribution Networks,* (pp. 1–31). Hershey, PA: IGI Global.
12. Sivaraman, P., and Sharmeela, C. (2020). Existing issues associated with electric distribution system. In Baseem Kahn, Hassan Haes Alhelou & Ghassan (Eds.), *Handbook of Research on New Solutions and Technologies in Electrical Distribution Networks,* (pp. 1–31). Hershey, PA: IGI Global.
13. Daniel Pinheiro Bernardon and Vinicius Jacques Garcia, *Smart Operation for Power Distribution Systems*, Springer, 2018.
14. Fabio Saccomanno, *Electric Power Systems: Analysis and Control,* 1st Edison, IEEE Press, 2003.
15. J. A. Clarke, *Energy Simulation in Building Design*, 2nd Edison, Butterworth – Heinemann, 2001.
16. Loi Lei Lai and Tze Fun Chan, *Distributed Generation: Induction and Permanent Magnet Generators*, IEEE press, 2007.
17. Turan Gonen, Electric *Power Distribution Engineering,* 3rd Edison, CRC Press, 2014.
18. William H. Kersting, *Distribution System Modelling and Analysis,* 4th Edison, CRC Press, 2018.
19. IEEE Std 18-2002, IEEE Standard for Shunt Power Capacitors.
20. Hemchandra Madhusudan Shertukde, *Distributed Photovoltaic Grid Transformers*, CRC Press, 2014.
21. Omid A Ardakanian, S. Keshav and Catherine Rosenberg, *Integration of Renewable Generation and Elastic Loads into Distribution Grids*, Springer, 2016.
22. Ali Emadi, Abdolhosein Nasiri and Stoyan B. Bekiarov, *Uninterruptible Power Supplies and Active Filters*, CRC Press, 2004.
23. IEEE Std 1015-2006, IEEE Recommended Practice for Applying Low Voltage Circuit Breakers Used in Industrial and Commercial Power Systems.
24. IS 2705-2 (1992): Current transformers, Part 2: Measuring current transformers
25. IS 2705-3 (1992): Current transformers, Part 3: Protective current transformers
26. IS 3156-3 (1992): Voltage transformers, Part 3: Protective voltage transformers
27. IS 3156-2 (1992): Voltage transformers, Part 2: Measuring voltage transformers
28. Sivaraman, P., and Sharmeela, C. (2020). Power Quality and its Characteristics. In Sanjeevikumar Padmanaban, C. Sharmeela, Jens Bo Holm-Nielsen, *Power Quality in Modern Power Systems*, Elsevier.
29. Sivaraman, P., and Sharmeela, C. (2020). Power System Harmonics. In Sanjeevikumar Padmanaban, C. Sharmeela, Jens Bo Holm-Nielsen, *Power Quality in Modern Power Systems*, Elsevier.

30. C. Sharmeela, Sivaraman, P., and S. Balaji (2020). Design of Hybrid DC Mini Grid for Educational Institution: Case Study, Lecture Notes in Electrical Engineering, 580, pp. 125-134.
31. P. Sivaraman and C. Sharmeela, (2020). IoT Based Battery Management System for Hybrid Electric Vehicle. In Chitra A, Sanjeevikumar Padmanaban, Jens Bo Holm-Nielsen and S. Himavathi, *Artificial Intelligent Techniques for Electric and Hybrid Electric Vehicles*, Scrivener Publishing.
32. P. Sivaraman, D. Gunapriya, K. Parthiban and S. Manimaran, "Hybrid Fuzzy PSO Algorithm for Dynamic Economic Load Dispatch", Journal of Theoretical and Applied Information Technology, Vol. 62, No.3, pp.794-799, April 2014.
33. P. Sivaraman, S. Manimaran, K. Parthiban and D. Gunapriya, "PSO Approach for Dynamic Economic Load Dispatch Problem", Int. Journal of Innovative Research in Science, Engineering and Technology, Vol. 3, No.4, pp.11905-11910, April 2014.
34. P. Sivaraman and C. Sharmeela, "Battery Energy Storage System Addressing the Power Quality Issue in Grid Connected Wind Energy Conversion System" 1st International conference on Large scale grid integration renewable energy in India, September, 2017, New Delhi, India.

# 2
# Transformers

## 2.1 Introduction

The Alternating Current (AC) systems are widely used over Direct Current (DC) systems because voltage in AC systems can be transferred from one voltage level to another voltage level. The device which is used to transfer the voltage from low voltage to high voltage and high voltage to low voltage is called a transformer [1]. It transforms the voltage level from one to another level at the same frequency. The typical Single Line Diagram (SLD) of power system with a step-up and step-down transformer is shown in Figure 2.1.

A transformer consisting of two windings (primary winding and secondary winding) which are electrically separated and connected through the magnetic circuit or core. The working principle of a transformer is based on Faraday's law of electromagnetic mutual Induction. It states that when two coils are identically coupled and if the current in one winding is changed uniformly then an Electro Magnetic Force (EMF) is induced in the other winding [2]. This EMF can drive a current. Figure 2.2 shows the typical configuration of a two-winding transformer.

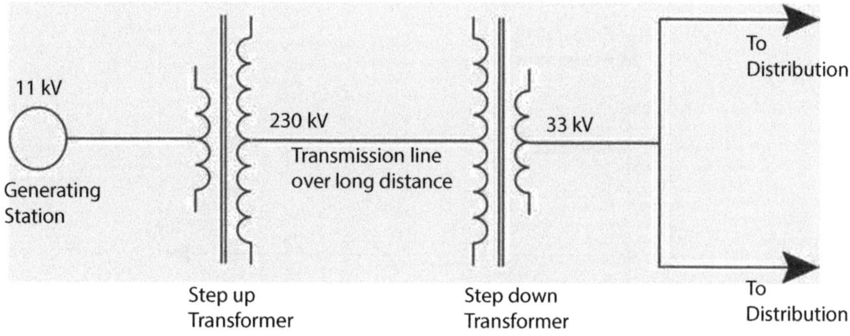


**Figure 2.1** Typical SLD of power system.

P. Sivaraman, C. Sharmeela, A. Thaiyal Nayagi, and R. Mahendran. Basic Electrical and Instrumentation Engineering, (79–124) © 2021 Scrivener Publishing LLC

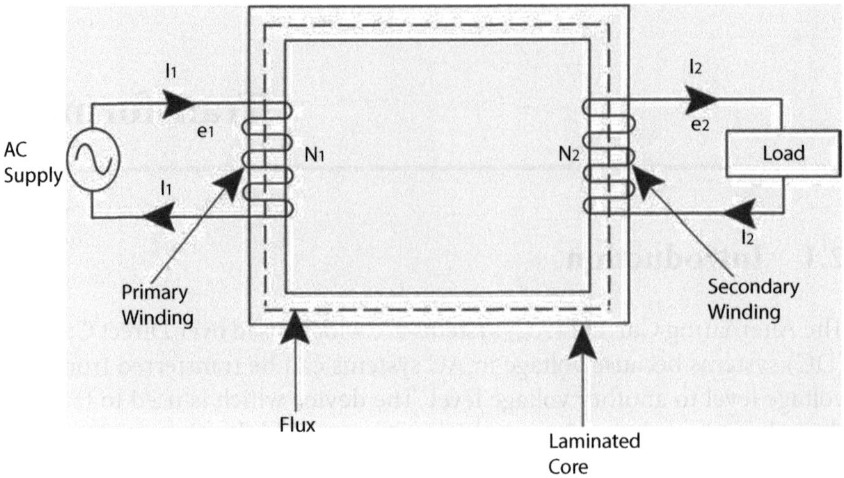


**Figure 2.2** Typical configuration of two-winding transformer.

Figure 2.3 shows the voltage and frequency of high voltage side of the transformer and Figure 2.4 shows the voltage and frequency of low voltage side of transformer.

Generally the winding connected with the power source is called primary or high voltage winding, and the number of turns of the winding is $N_1$. The winding connected to the load is called secondary or low voltage winding and number of turns of the winding is $N_2$. When high voltage winding is excited by an alternating voltage, it creates the magnetic flux

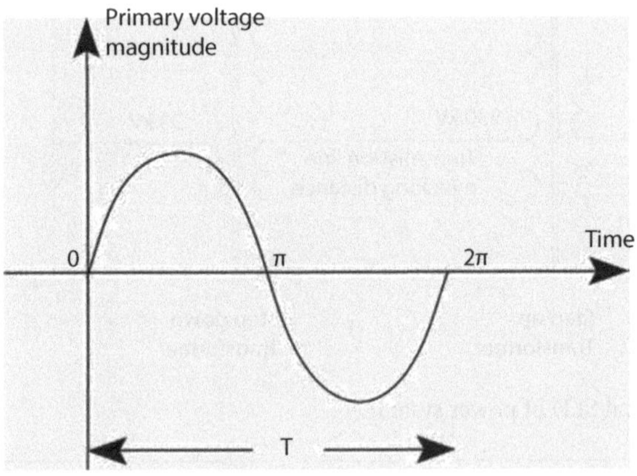


**Figure 2.3** Waveform of high voltage side of two-winding transformer.

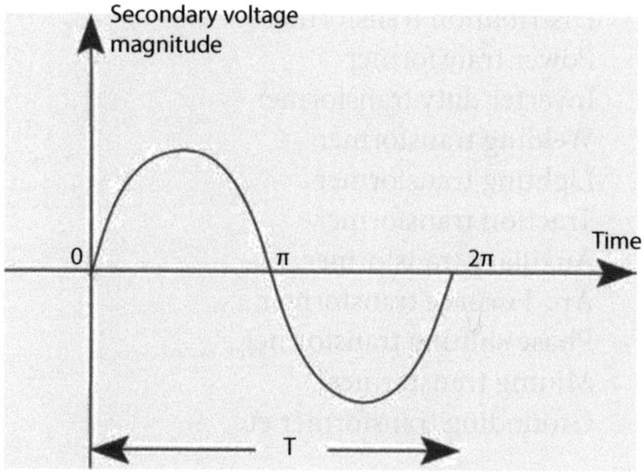


**Figure 2.4** Waveform of low voltage side of two-winding transformer.

and it circulates around the core of the transformer. This alternating flux created by high voltage winding links with the low voltage winding.

According to Faraday's law of electromagnetic induction, voltage is induced in low voltage winding based on mutual inductance.

Based on the construction, a transformer is classified into two types:

I. Core-type transformer
II. Shell-type transformer

Based on the voltage output, a transformer is classified into four types:

I. Step-up transformer
II. Step-down transformer
III. Auto-transformer
IV. 1:1 ratio transformer

Based on the duty they are working, a transformer is classified into two types:

I. Power transformer
II. Instrument transformer
  a) Current transformer
  b) Voltage transformer or potential transformer

Based on the application, a transformer is classified into the following types [2, 16]:

I. Distribution transformer
II. Power transformer
III. Inverter duty transformer
IV. Welding transformer
V. Lighting transformer
VI. Traction transformer
VII. Auxiliary transformer
VIII. Arc-Furnace transformer
IX. Phase shifting transformer
X. Mining transformer
XI. Grounding transformer etc.

Based on the number of phases, a transformer is classified into two types:

I. Single-phase transformer
II. Three-phase transformer

Based on the number of windings, a transformer is classified into the following types:

I. Single-winding transformer (auto transformer)
II. Two-winding transformer
III. Three-winding transformer
IV. Five-winding transformer

## 2.2 Transformer Magnetics

The magnetic circuit acts as the basis for the working of a transformer. The material for the core is chosen such that it has high permeability at rated frequency (usually measured at 50Hz or 60Hz for power transformers) so that the net reluctance is reduced during operation. The typical materials used for the core are as follows:

1. Hot Rolled Grain Oriented (HRGO) silicon steel: Max. Flux Density = 1.2 Tesla, Typical Design Flux Density = 0.8 - 1 Tesla.

2. Cold Rolled Grain Oriented (CRGO) silicon steel: Hi-B lazer scribed or mechanically scribed (Max. Flux Density = 2.0 Tesla, Typical Design Flux Density = 1.4 - 1.8 Tesla)

3. Amorphous core: Max. Flux Density = 2.6 Tesla, Typical Design Flux Density = 2.0 - 2.4 Tesla

The relative permeability of a core is usually less than 4000 for HRGO, around 4000 for CRGO and higher for Amorphous materials. The core of a three-phase transformer is shown in Figure 2.5.

Losses in the magnetic circuit: In a practical high-power transformer, the magnetic losses are not just confined to the core, but can occur anywhere in and around the flux carrying path. Between 95 and 97% of the magnetic circuit losses occur inside the core, the remaining 3-5% are contributed by the eddy current in the structural parts of the transformer. These currents are usually induced in the turret, tank and the condenser foils of the bushing. These losses are referred as stray magnetic circuit losses.

For transformers whose MVA rating is less than 50MVA, the stray losses become negligible, the core losses are the concern. Core losses mainly arise as a result of material hysteresis and eddy currents. The core losses are measured in Watts per Kg.

i. Hysteresis loss:

The relationship between applied magnetic field intensity (H) and the corresponding magnetic flux density (B) is referred to as hysteresis. In other words, it is a measure of magnetic polarization in the direction of the applied magnetic field. Since the polarization phenomena is inherently

**Figure 2.5** Core of three-phase power transformer (Courtesy: Andrew Yule).

nonlinear, the magnetic field intensity VS magnetic flux density (B-H) curve is also nonlinear in nature [3], as shown in Figure 2.6. The nonlinearity can be reduced by good choice of core material. The material which has the least loop area is regarded as the best material.

Remanence Magnetization (BR) is the residual flux density that the material retains even after removal of the initially applied magnetic field intensity (H). For a good core, the remanence magnetization is less than 60% of the saturation flux density.

An empirical relationship between the operating frequency, the magnetic flux density and the hysteresis loss is given in equation 2.1.

$$P_h = K_1 \times F \times B_p^n \cdots \tag{2.1}$$

Where,

$K_1$ is the proportionality constant
F is the frequency
$B_p$ is the peak of flux density
n is the stienmetz constant (n = 1.6 for CRGO and n = 2 for HRGO)

ii. Eddy current losses:
The eddy current losses occur due to induced currents in the core along the magnetic flux path. The magnitude of induced currents depends on the core material resistivity, frequency (F) and the flux density (B) [4]. The values of resistivity ranges from 450-550X$10^9$ Ωm depending upon the grade of material used. The expression for power loss due to eddy current is given in equation 2.2.

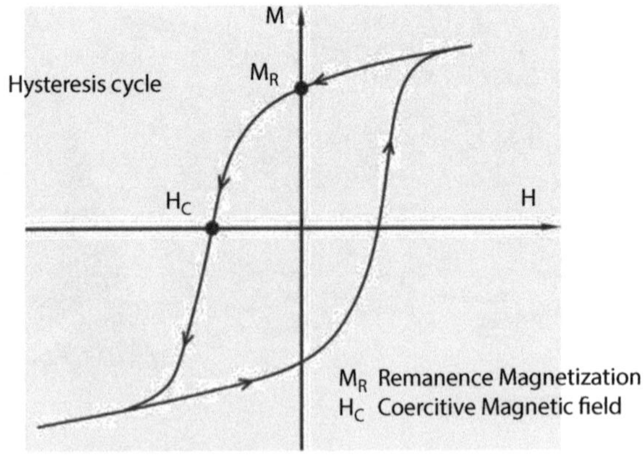


**Figure 2.6** Relationship of B-H curve.

$$P_e = K_2 \times F^2 \times B^2 \times t^2 \tag{2.2}$$

Where,

$K_2$ is the proportionality constant (derived empirically for each material)
t is the thickness of lamination

Eddy current losses are usually high in magnitude. In order to reduce this eddy current loss, the core is generally segmented and laminated as shown in Figure 2.7. Lamination prevents constrictions of the path of induced current flow thus increasing the effective resistance along the path of the eddy currents as shown in Figure 2.8.

## 2.3 Construction of Transformer

The main parts of any transformers are:

(i) Primary or High voltage winding
(ii) Secondary or Low voltage winding
(iii) Core

The primary or High voltage winding of the transformer is generally made up of a copper or aluminium conductor. This winding is connected to the source side for a step-down transformer.

**Figure 2.7** CGL power transformer laminated core (Courtesy: CGL power transformer).

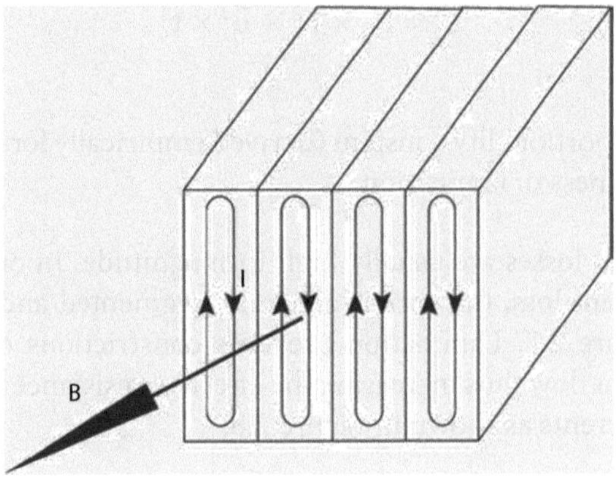


**Figure 2.8** Path of eddy current.

The secondary or Low voltage winding of the transformer is generally made up of a copper or aluminium conductor. This winding is connected to the load side for a step-down transformer.

Transformer cores are made up of magnetic material, generally silicon steel. Both primary and secondary windings are wound on the transformer core to achieve electrical isolation between both the windings and magnetically coupled. This core is made up of laminated stampings in order to reduce the eddy current loss [2]. The typical thickness of the lamination is between 0.35 mm to 0.5 mm. Based on the core arrangement, transformer cores are classified into two types:

(i) Core-type transformer
(ii) Shell-type transformer

**A. Core-type Transformer:**
The core-type transformer has both high voltage winding and low voltage winding wound in a common core to form a single magnetic circuit. It is shown in Figure 2.9.

It has two L shape stampings as shown in Figure 2.10.

Generally, core-type transformer windings are wound with circular cylindrical coil in helical layers with different layers insulated from each other by paper or mica. Both high voltage and low voltage windings are wound on both limbs of the core. The core is made up of a large number of

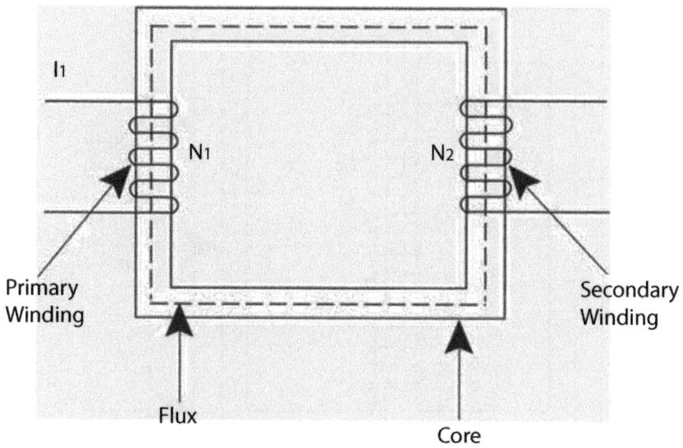


**Figure 2.9** Core-type transformer.

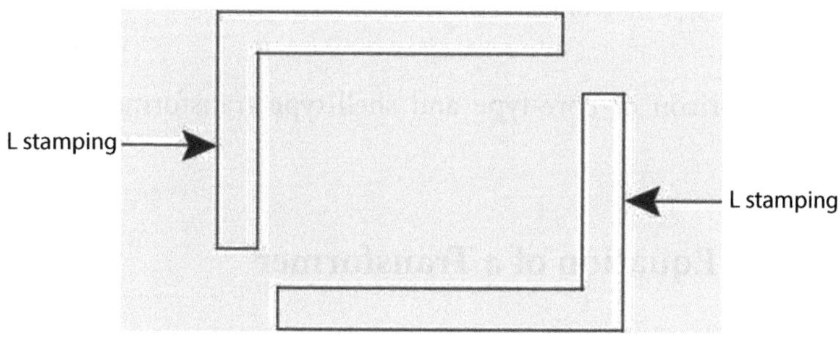


**Figure 2.10** L shape stamping.

thin laminations. Insulation layers are used to isolate the electrical contact between the windings and the core.

### B. Shell-type Transformer

The shell-type transformer has both high voltage winding and low voltage winding and is wound in a common core to form a double magnetic circuit. It is shown in Figure 2.11.

This transformer core has three limbs. Both the high voltage and low voltage windings are wound on the central limb. The windings are generally wound as multilayer type. The core is laminated so that all the joints at alternate layers are staggered. This is done to avoid the narrow air gap at the joints; such joints are called overlapped or imbricated joints. The core is made up of E and I shape stampings as shown in Figure 2.12.

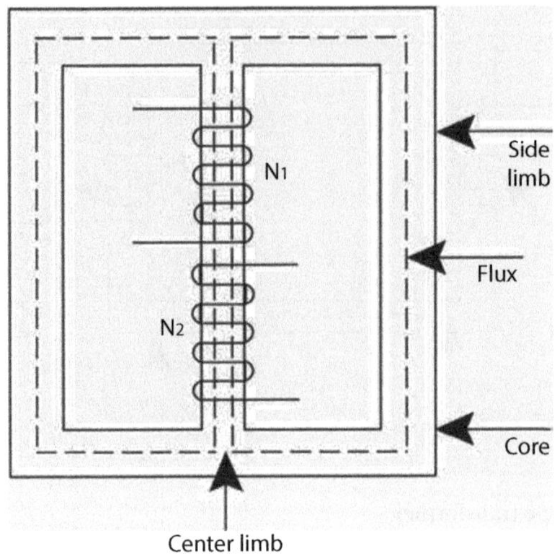


**Figure 2.11** Core-type transformer.

The comparison of core-type and shell-type transformer is listed in Table 2.1.

## 2.4 EMF Equation of a Transformer

The EMF equation of a transformer is used to find the induced voltage at low voltage side of the step-down transformer and high voltage side of the step-up transformer. The typical step-down transformer is shown in Figure 2.13.

Where

$V_1$ is supply voltage to primary winding (V)

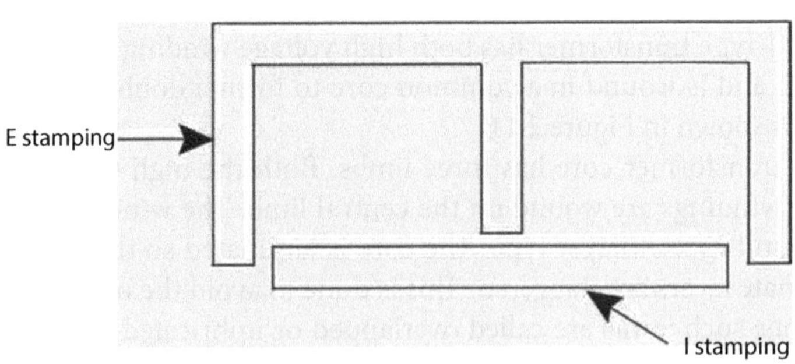


**Figure 2.12** E & I shape stamping.

**Table 2.1** Comparison of core-type and shell-type transformer.

| S. No. | Core type | Shell type |
|---|---|---|
| 1 | It has single magnetic circuit | It has double magnetic circuit |
| 2 | The core has two limbs | The core has three limbs |
| 3 | The cylindrical coils are used | The multilayer disc or sandwich type coils are used |
| 4 | The windings are uniformly distributed on two limbs, hence natural cooling effective | The natural cooling does not exist as per windings are surrounded by the core |
| 5 | The coils can be easily removed for maintenance | The coils cannot be removed easily |
| 6 | It is preferred for low voltage transformer | It is preferred for high voltage transformer |

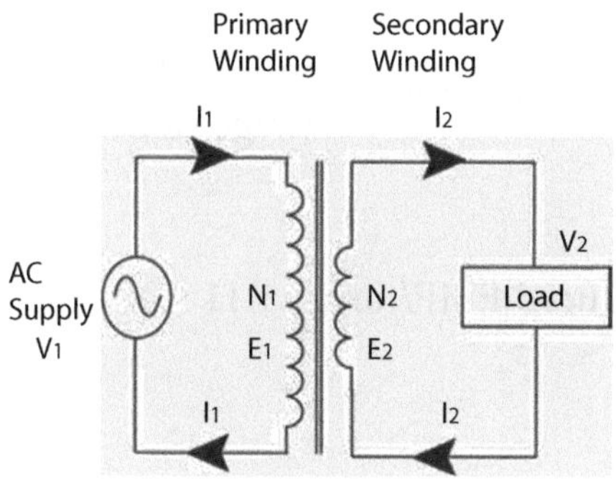


**Figure 2.13** Typical step-down transformer.

$V_2$ is Terminal voltage across the secondary (V)
$N_1$ is Number of turns in primary winding
$N_2$ is Number of turns in secondary winding
$I_1$ is Rated full load primary current (A)
$I_2$ is Rated full load secondary current (A)
$E_1$ is Induced EMF in primary winding (V)
$E_2$ is Induced EMF in secondary winding (V)
F is Frequency (Hz)
A is Core area ($m^2$)
$B_m$ is Maximum value of flux density in the core ($Wb/m^2$)

$\Phi_m$ Maximum value of flux in the core (Wb)

The flux established on the transformer core is alternating or varying as applied input voltage is alternating with respect to time. It is shown in Figure 2.14.

From Figure 2.14, it is evident that flux is varying with respect to time and attains the maximum value ($\phi_m$) at time period (T/4) of every cycle.

From equation 1.8, $T = \frac{1}{F}$

Where,

T is time (ms)

F is frequency (Hz)

Average rate of flux change = $\frac{\Phi_m}{1/4F}$ Wb/seconds

As per Faraday's law of electromagnetic induction, the average value of EMF induced/turn = $4F \times \Phi_m$ (V).

For sinusoidal waveform,

$$\text{Form factor} = \frac{\text{RMS value}}{\text{Average value}} = 1.11$$

$$\text{RMS value of induced EMF/turn} = 1.11 \times 4F \times \Phi_m = 4.44\,F\Phi_m \text{ (V)} \quad (2.3)$$

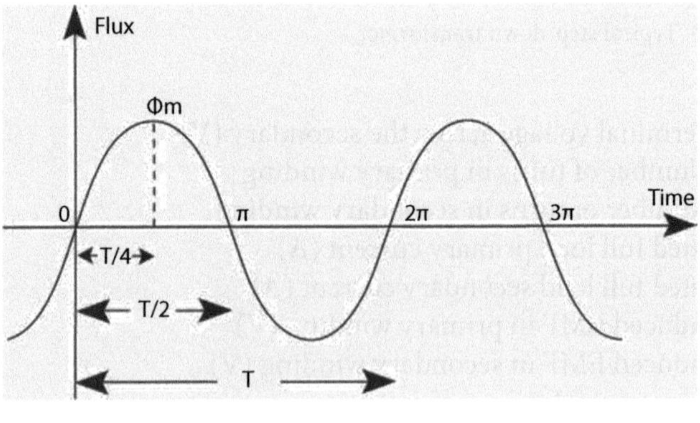


**Figure 2.14** Flux on the transformer core.

RMS value of induced EMF in the primary winding (entire winding)

$$E_1 = 4.44\ F\Phi_m \times N_1$$

$$\Phi_m = B_m \times A$$

$$\text{So, } E_1 = 4.44 \times F \times B_m \times A \times N_1\ (V) \qquad (2.4)$$

Similarly, RMS value of induced EMF in the secondary winding (entire winding)

$$E_2 = 4.44\ F\Phi_m \times N_2\ (V) \text{ or}$$

$$E_2 = 4.44 \times F \times B_m \times A \times N_2\ (V) \qquad (2.5)$$

## 2.5 Ideal Transformer

The definition for an ideal transformer is an imaginary transformer which has 100% efficiency and doesn't have any losses (both core and copper losses) [2].

The following assumptions are considered for an ideal transformer:

i. The ideal transformer has zero winding resistance. This means, Ohmic losses and voltage drops offered by resistance of the transformer are ignored.
ii. The transformer core has infinite permeability. So it requires zero MMF to create flux in the core.
iii. The leakage flux is zero, i.e., flux created by primary winding is fully linked in secondary winding.
iv. The transformer core losses are zero.

The ideal transformer is shown in Figure 2.15.

The supply voltage is $V_1$ and its secondary no load current $I_2$ is zero. The current drawn by primary winding $I_1$ is necessary to produce flux in the core. As the core is made up of magnetic material, the current drawn by the primary winding is called magnetizing current and denoted as $I_m$. The magnetizing current $I_m$ is very small and it lags the supply voltage ($V_1$) by 90°. The flux created by magnetizing current $I_m$ is Φ and which is in phase with $I_m$. The flux links with both the windings and producing the induced EMF ($E_1$ in primary winding and $E_2$ in secondary winding). According to

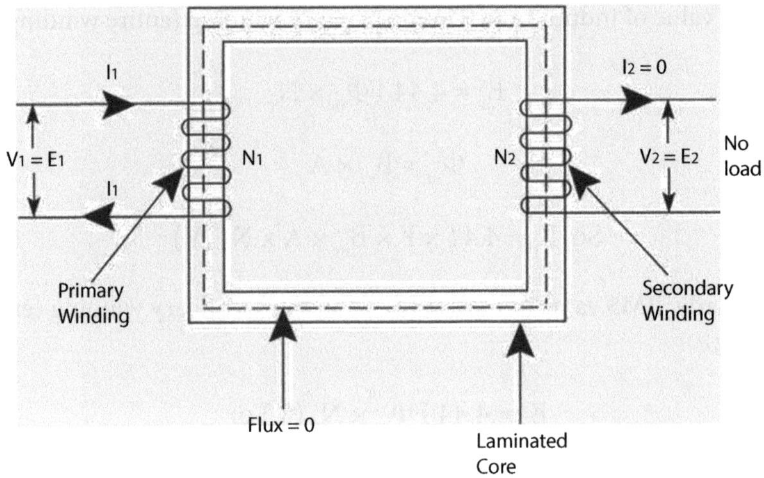


**Figure 2.15** The ideal transformer.

Lenz law, the induced EMF is opposite to the supply voltage $V_1$. That is $E_1$ is opposite with $V_1$ but equal in magnitude.

The induced EMF, $E_2$ also opposite to supply voltage ($V_1$) and its magnitude depends on the secondary winding turn ($N_2$). The direction of induced EMF, $E_1$ and $E_2$ are in phase. Figure 2.16 shows the voltage – current relationship of an ideal transformer.

From Figure 2.16, $E_1$ lags current $I_m$ by 90°. No load input current $I_o$ has two components. A purely reactive component ($I_m$) is called magnetizing

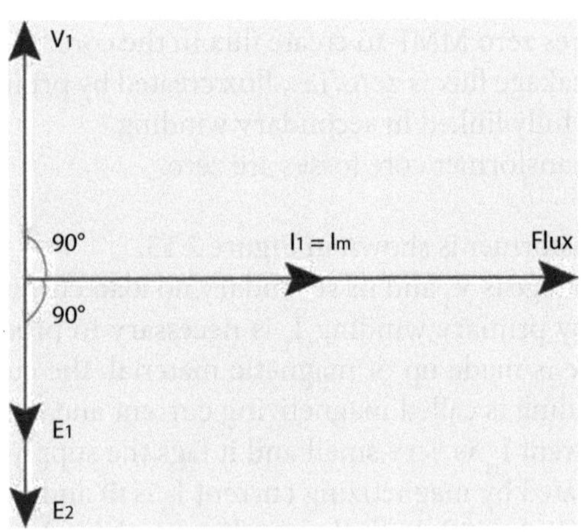


**Figure 2.16** Voltage – current relationship of an ideal transformer.

component of no load current to produce the flux. This component is also called lose less component. A real or active component ($I_c$) that supplies the total losses under no load condition is called power component. This component is called Wattful component or core loss component of $I_o$.

$$\overline{I_o} = \overline{I_m} + \overline{I_c}$$

$$I_m = I_o \text{ Sin } \phi_o$$

$$I_c = I_o \text{ Cos } \phi_o$$

$$I_o = \sqrt{I_m^2 + I_c^2} \tag{2.6}$$

$$W_o = V_1 I_o \text{Cos } \phi_o = V_L I_c \tag{2.7}$$

$$W_o = V_1 I_o \text{Cos } \phi_o \tag{2.8}$$

$$\text{Pi} = \text{Iron Loss}$$

In a real transformer, the core has finite permeability and in order to establish flux in the core, primary winding will draw the current which has the magnetizing current component from the source along with the load current. Assuming the core to be linear, let the core reluctance be 'R'. The magnetizing current is given to be

$$I_m = \frac{R\phi}{N_1} A \tag{2.9}$$

$$\therefore R = \frac{\text{mmf}}{\phi} \Rightarrow \frac{NI_m}{\phi}$$

Where,

R is reluctance

The phasor diagram is shown in Figure 2.17 and the circuit diagram is shown in Figure 2.18.

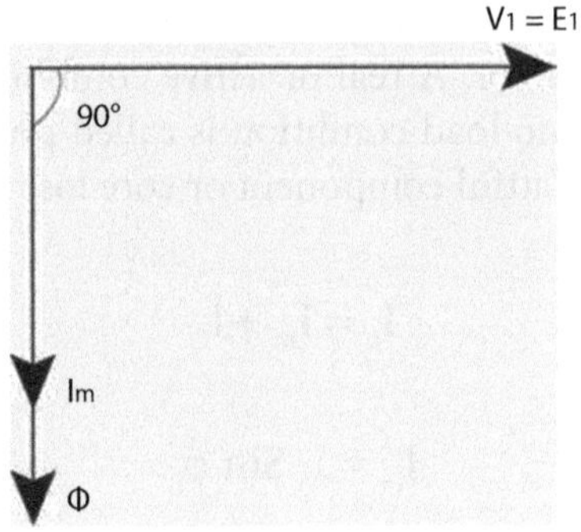


**Figure 2.17** Phasor diagram.

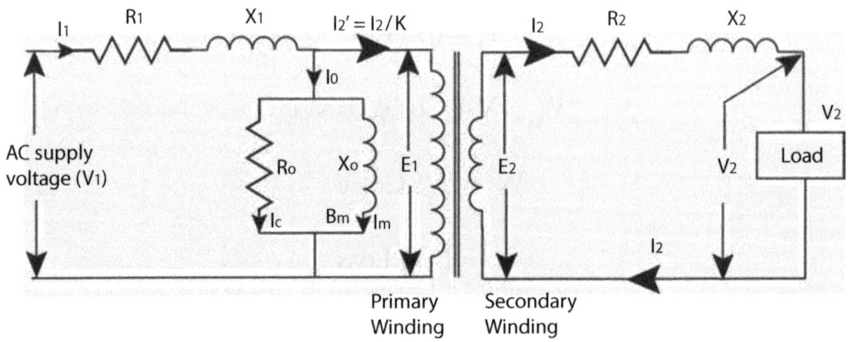


**Figure 2.18** Circuit diagram.

$$\overline{\mathrm{Im}} = \frac{\overline{V1}}{j\mathrm{Xm}} = -j\mathrm{Bm}\overline{V1} \tag{2.10}$$

$$B_m = \frac{1}{X_m} \Rightarrow \frac{R}{WN_1^2} \tag{2.11}$$

In a circuit model, reactance ($X_m$) draws the magnetizing current from the primary voltage source. A core also has core loss because it carries alternating flux. It can be modeled as a reactance ($R_i$) or conductance ($G_i$) across the primary voltage source. This is accruable representation for constant frequency operation and power loss is $G_iV_1^{\,2}$ while the actual core loss has two components.

i. Eddy current loss is directly proportional to $V_1^2(\phi_{max}\alpha\ V_1)$
ii. Hysteresis loss proportional to $V_1^{\ 1.6}$

The net current drawn by the primary winding to create core flux is

$$\overline{Io} = \overline{\mathrm{Im}} + \overline{Ic} \tag{2.12}$$

$$\text{Magnetizing Current} = \overline{\mathrm{Im}} = -jBm\overline{V1}$$

$$\text{Core (or) Iron Loss Coefficient} = \overline{\mathrm{Ic}} = Gi\overline{V1}$$

The magnitude $I_c$ is smaller than $I_m$ and the angle $\theta o$ is close to 90°. It is also clear that under no load $\left(I_2{}' = \frac{I_2}{K} = 0\right)$ the transformer primary can draw only the exciting current from the source which is therefore synchronous with the term no load current ($I_0$). The circuit has assumptions that windings resistance is less, leakage flux is negligible and the core is linear. In real core, in order to produce the sinusoidal core flux, the exciting current $I_0$ is periodic but non-sinusoidal with a predominant third harmonics [17], [18]. $\bar{I}_1 = \bar{I}_0 + \bar{I}_2{}'$ magnetizing current or exciting current is predominantly reactive. Therefore it has to be as low as possible. That's why transformer cores are constructed with high permeability of steel. The magnetizing current drawn by the transformer is in the range of 5 to 7.5% of the rated current.

## 2.6 Transformation Ratio (K)

For an ideal transformer, transformation ratio (K) can be calculated as follows [2]:

$$V_1 = E_1$$

$$V_2 = E_2$$

$$V_1 I_1 = V_2 I_2$$

$$\frac{V_2}{V_1} = \frac{I_1}{I_2}$$

$$\frac{E_2}{E_1} = \frac{I_1}{I_2}$$

$$\frac{E_2}{E_1} = \frac{N_2}{N_1}$$

$$\frac{E_2}{E_1} = \frac{N_2}{N_1} = \frac{I_1}{I_2} = K \quad (2.13)$$

If $N_2 > N_1$. i.e $K > 1$, then the transformer is called a step-up transformer

If $N_2 < N_1$. i.e $K < 1$, then the transformer is called a step-down transformer

$$\text{Voltage ratio} = \frac{E_2}{E_1} = K$$

$$\text{Current ratio} = \frac{I_2}{I_1} = \frac{1}{K}$$

## 2.7 Circuit Model or Equivalent Circuit of Transformer

The equivalent circuit of the transformer has a combination of fixed and variable resistance and reactance. The resistance and reactance simulates the performance and working of the machine. For a transformer, no load primary current $I_o$ has two components as follows:

Magnetizing current component = $I_m = I_o \text{ Sin } \Phi_o$

Real or active current component = $I_c = I_o \text{ Cos } \Phi_o$

The no load equivalent circuit is shown in Figure 2.19.

The magnetizing current $I_m$ produces the flux and the flux is assumed to flow through reactance $X_o$ is called no load reactance. The $I_c$ is active component representing core loss and assumed to flow though the

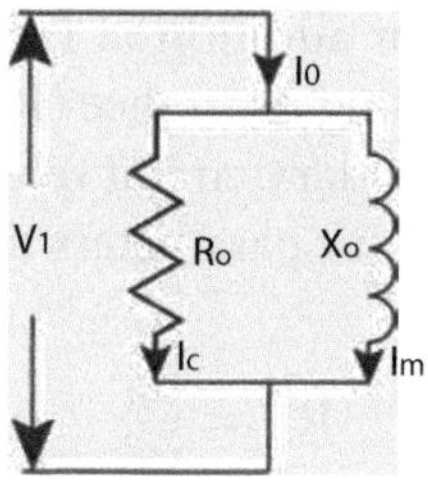


**Figure 2.19** No load equivalent circuit.

resistance $R_o$. This circuit consist of $R_o$ and $X_o$ in parallel is called exciting circuit.

$$R_o = \frac{V_1}{I_c} \text{ and } X_o = \frac{V_1}{I_m}$$

The equivalent circuit of a transformer under load in secondary side is shown in Figure 2.20.

When the load connected to the secondary side of the transformer, the current $I_2$ flows in secondary side cause the voltage drop across the resistance ($R_2$) and reactance ($X_2$). Due to secondary load current $I_2$, primary winding draws an additional current $I_2'$.

$$I_2' = \frac{I_2}{K} \tag{2.14}$$

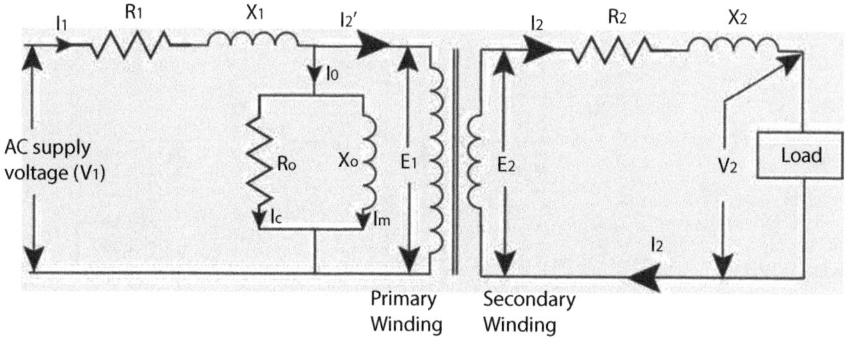


**Figure 2.20** Equivalent circuit of a transformer under load in secondary side.

The current $I_1$ is the phasor addition of $I_o$ and $I_2'$, it causes the voltage drop across the primary winding resistance ($R_1$) and reactance ($X_1$).

Figure 2.21 shows the secondary circuit referred to the primary circuit.

So, transferring the secondary parameters to primary.

$$R_2' = \frac{R_2}{K^2}$$

$$X_2' = \frac{X_2}{K^2}$$

$$Z_2' = \frac{Z_2}{K^2}$$

While

$$E_2' = \frac{E_2}{K}$$

$$I_2' = KI_2$$

Where,

$$K = \frac{N_2}{N_1}$$

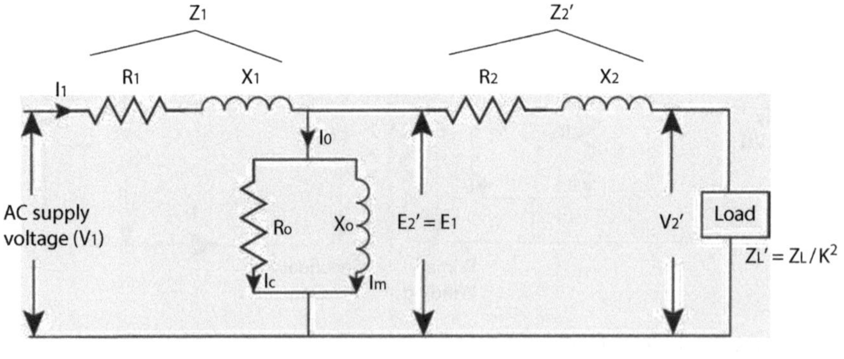


**Figure 2.21** Secondary circuit referred to the primary circuit.

The points to remember while transferring the values
Low voltage winding → High current → Low impedance
High voltage winding → Low current → High Impedance

Figure 2.22 shows the primary circuit referred to the secondary circuit. So, transferring the secondary parameters to primary

$$R_1' = K^2 R_1$$

$$X_1' = K^2 X_1$$

$$Z_1' = K^2 Z_1$$

$$E_1' = K E_1$$

$$I_1' = \frac{I_1}{K}$$

$$I_0' = \frac{I_0}{K}$$

In order to get approximate equivalent circuit of a transformer, shift the no load branch containing $R_o$ and $X_o$ to the left of $R_1$ and $X_1$. By doing this, error is created that the drop across $R_1$ and $X_1$ due to $I_o$ is neglected. Hence such an equivalent circuit is called approximate equivalent circuit. It is shown in Figure 2.23.

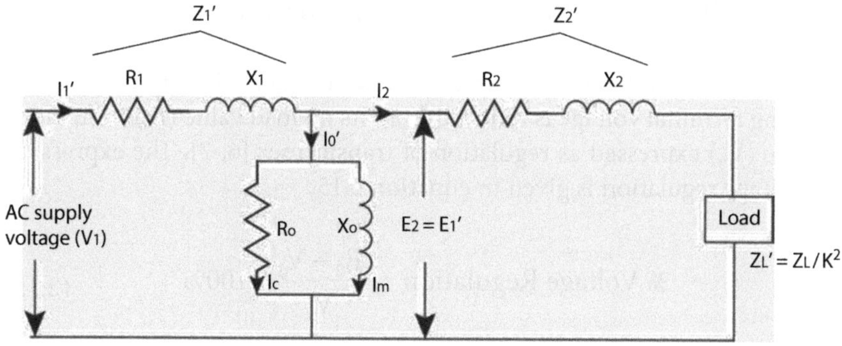


**Figure 2.22** Primary circuit referred to the secondary circuit.

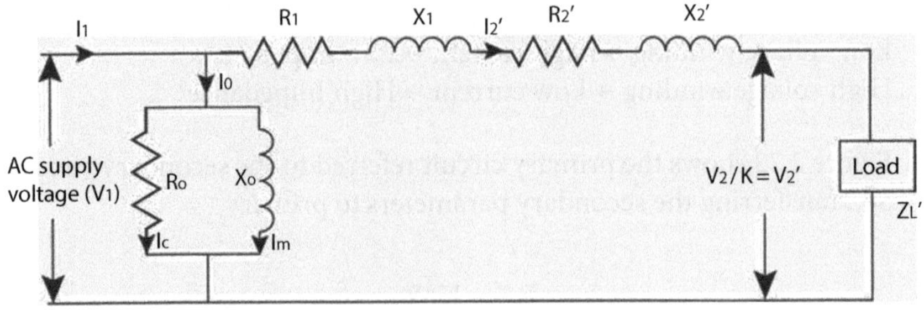


**Figure 2.23** Approximate equivalent circuit.

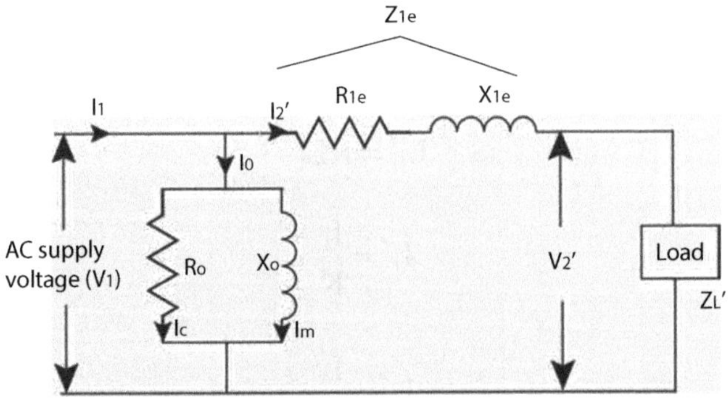


**Figure 2.24** Equivalent circuit.

In Figure 2.23, $R_1$ and $R_2$' can be combined to get equivalent resistance referred to primary $R_{1e}$. Similarly $X_1$ and $X_2$' can be combined to get $X_{1e}$ and equivalent circuit is simplified as shown in Figure 2.24.

## 2.8 Voltage Regulation of Transformer

The impedance offered by primary and secondary windings, secondary winding terminal voltage is reduced from its no load value ($E_2$) to full load voltage ($V_2$) expressed as regulation of transformer [6, 7]. The expression for voltage regulation is given in equation 2.15.

$$\%\ \text{Voltage Regulation} = \frac{E_2 - V_2}{V_2} \times 100\% \tag{2.15}$$

Where,

$E_2$ is secondary terminal voltage on no load

$V_2$ is secondary terminal voltage on load

The voltage-current characteristic of voltage regulation of the transformer is shown in Figure 2.25.

For lagging power factor load, voltage regulation is given equation 2.16.

$$\%R = \frac{E_2 - V_2}{V_2} \times 100 = \frac{I_2[R_{2e}\cos\phi_2 + X_{2e}\sin\phi_2]}{V_2} \times 100 \quad (2.16)$$

For leading power factor load, voltage regulation is given equation 2.17.

$$\%R = \frac{E_2 - V_2}{V_2} \times 100 = \frac{I_2[R_{2e}\cos\phi_2 - X_{2e}\sin\phi_2]}{V_2} \times 100 \quad (2.17)$$

## 2.9 Name Plate Rating

The transformer's defined electrical parameters like primary and secondary voltage, primary and secondary current, power rating, frequency, impedance, ambient temperature, etc., are specified in the name plate rating [7, 8]. The typical name plate rating of the transformer is shown in Figure 2.26.

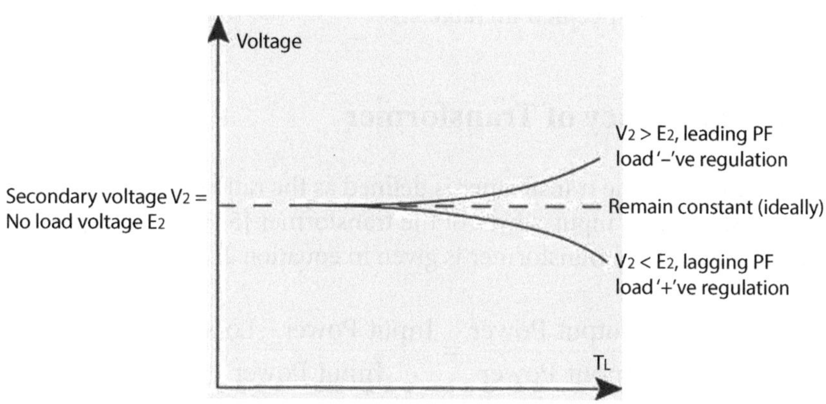


**Figure 2.25** Voltage regulation of a transformer.

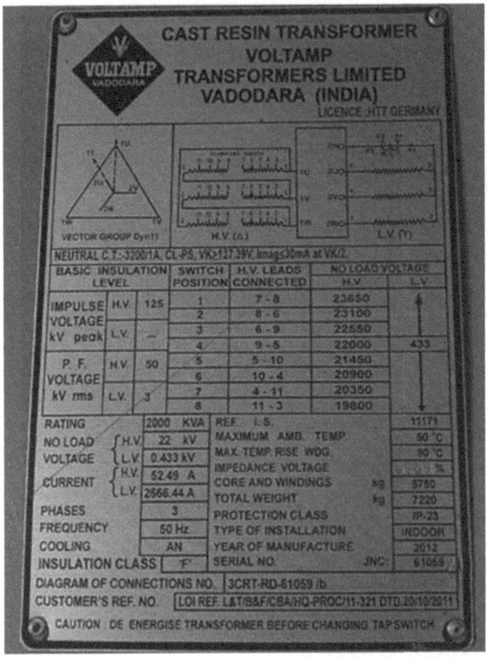


**Figure 2.26** Typical name plate rating of a transformer (Courtesy: Voltamp transformer).

The name plate details of the transformer shown in Figure 2.26 are listed in Table 2.2.

These values are called the name plate rating of the transformer. It means, at maximum of 50°C ambient temperature, this transformer has the parameters as specified in Table 2.2.

## 2.10 Efficiency of Transformer

The efficiency of the transformer is defined as the ratio of output power of the transformer to input power of the transformer [5, 15]. The expression for efficiency of the transformer is given in equation 2.18.

$$\eta = \frac{\text{Output Power}}{\text{Input Power}} = \frac{\text{Input Power} - \text{Losses}}{\text{Input Power}} (\%) \tag{2.18}$$

Due to the losses in a transformer, the output of a transformer is less than the input power.

**Table 2.2** Typical name plate rating of a transformer.

| S. No. | Parameter | Specification |
|---|---|---|
| 1 | Power rating (kVA) | 2000 |
| 2 | Primary voltage (kV) | 22 |
| 3 | Secondary voltage (kV) | 0.433 |
| 4 | Primary current (A) | 52.49 |
| 5 | Secondary current (A) | 2666.44 |
| 6 | Frequency (Hz) | 50 |
| 7 | Maximum ambient temperature (deg. Cel) | 50 |
| 8 | Impedance voltage (%) | 6.68 |
| 9 | Type of cooling | AN |
| 10 | Insulation class | F |

Output power = Power input – Total Losses
Input power = Power output + Total Losses
Output power of the transformer is $V_2I_2Cos\phi$
Where,
$Cos\phi$ is load power factor

The transformer supplies full load current $I_2$ with the terminal voltage $V_2$. The copper loss of the transformer is given in equation 2.19.

$$\text{Full load copper loss (Pcu)} = I_2^2R_{2e} \tag{2.19}$$

$$\eta = \frac{V_2I_2Cos\phi_2}{V_2I_2Cos\phi_2 + P_i + I_2^2R_{2e}}$$

$V_2I_2$ = VA rating of transformer

$$\eta = \frac{(\text{VA rating}) \times Cos\phi}{(\text{VA rating}) \times Cos\phi + P_i + I_2^2R_{2e}} \times 100\% \tag{2.20}$$

## 2.11 Three-Phase Transformer

The generation of an electrical power is usually three phase at medium voltage between 3.3 kV to 22 kV and transmission of electrical power is also at medium or high voltage like 110 kV, 132 kV or 400 kV. The generated voltages between 3.3 kV to 22 kV is stepped up to higher voltage like 110 kV, 132 kV or 400 kV for transmission purpose and stepped down to lower voltage like 33 kV, 22 kV, 11 kV or 6.6 kV, etc., for distribution purpose by means of a three-phase transformer [13, 14]. Three-phase transformers are most suitable for economical transmission, distribution and utilization of electrical energy.

**A. Principle of Operation**

The three cores are arranged at 120° to each other. The primary windings are wound on the cores and connected to the three-phase supply as shown in Figure 2.27. The primary current in all three phases are $I_R$, $I_Y$, $I_B$ and it produces the flux $\Phi_R$, $\Phi_Y$, $\Phi_B$ in the individual core. The common leg of the core, i.e., center leg formed, carries the sum of all three fluxes.

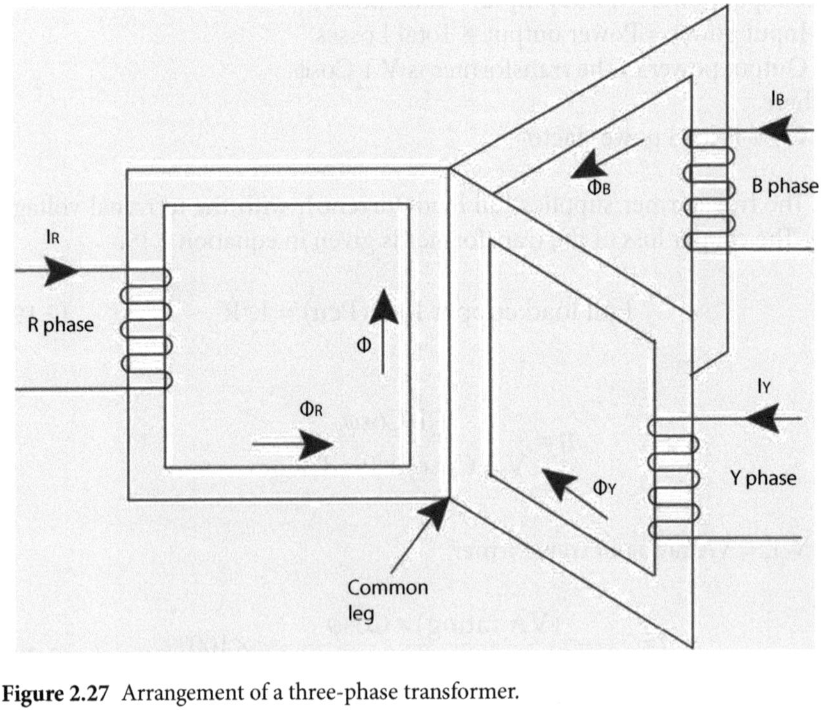


**Figure 2.27** Arrangement of a three-phase transformer.

At any instant in a three-phase system, vector sum of all three phase current is $I_R + I_Y + I_B = 0$, hence the sum of the three flux is also zero at any instant.

**B. Three-Phase Transformer Connections**

Based on the connection of both primary and secondary windings, three-phase transformer windings can be any of the following combinations of Star or in Delta.

i. Star - Star (Y - Y) connection
ii. Delta - Delta (Δ – Δ) connection
iii. Star - Delta (Y – Δ) connection
iv. Delta - Star (Δ – Y) connection
v. Open Delta or V connection
vi. Slot or T-T connection

I. Star - Star (Y - Y) connection

The star – star connection of primary and secondary winding of a three-phase transformer is shown in Figure 2.28.

The line voltage ($V_{L1}$) on the primary side, then phase voltage on primary side is $V_{ph1}$. The expression for $V_{ph1}$ is given equation 2.21.

$$V_{ph1} = \frac{V_{L1}}{\sqrt{3}} \tag{2.21}$$

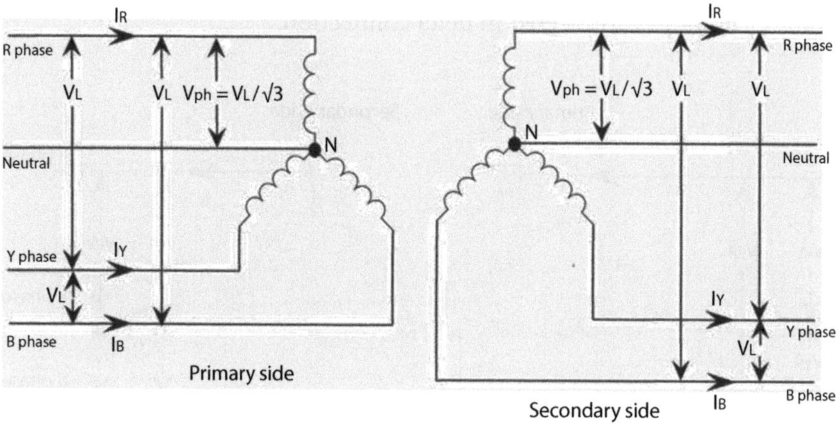


**Figure 2.28** Star – Star configuration of three-phase transformer.

K is turns ratio or transformation ratio, then phase voltage on secondary side is $V_{ph2}$. The expression for $V_{ph2}$ is given equation 2.22.

$$V_{ph2} = \frac{V_{L2}}{\sqrt{3}} \tag{2.22}$$

$$K = \frac{V_{ph2}}{V_{ph1}}$$

ii. Delta - Delta (Δ – Δ) connection

The delta – delta connection of primary and secondary winding of a three-phase transformer is shown in Figure 2.29.

$V_{L1}$, $V_{ph1}$ is line and phase voltage on primary side

$V_{L2}$, $V_{ph2}$ is line and phase voltage on secondary side

When the primary and secondary windings are delta connected, line voltage is equal to phase voltage.

$$V_{L1} = V_{ph1} \tag{2.23}$$

$$V_{ph2} = KV_{ph1} \tag{2.24}$$

$$V_{L2} = V_{ph2} = KV_{L1} \tag{2.25}$$

iii. Star - Delta (Y – Δ) connection

The Figure 2.30 shows the primary winding is connected in star and secondary winding is connected in delta connection.

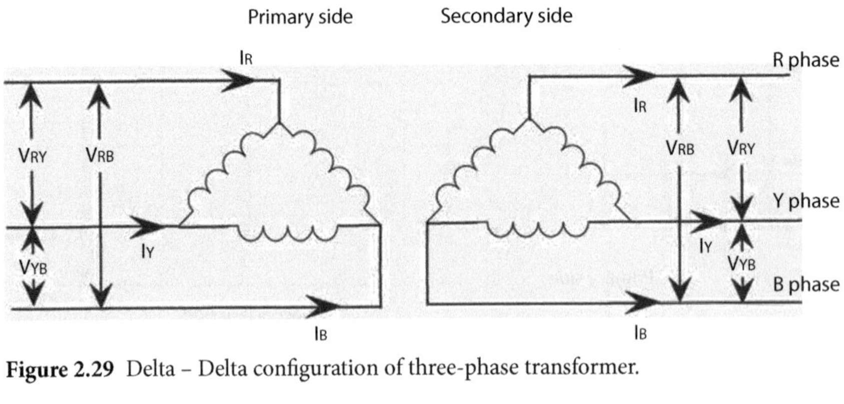


**Figure 2.29** Delta – Delta configuration of three-phase transformer.

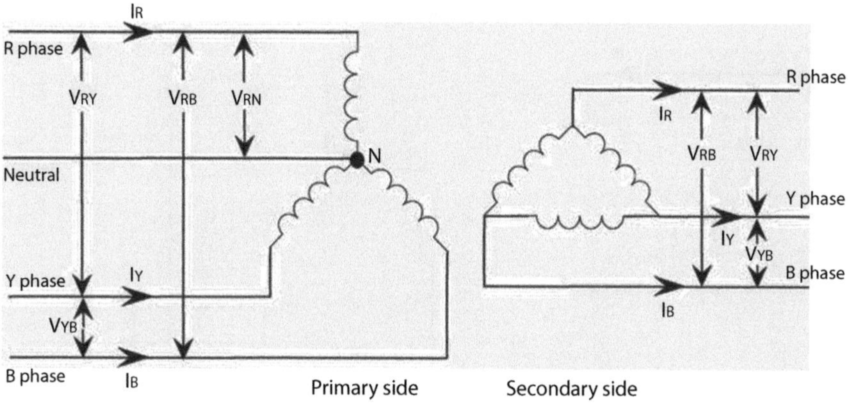


**Figure 2.30** Star – Delta configuration of three-phase transformer.

$V_{L1}$, $V_{ph1}$ is line and phase voltage on primary side
$V_{L2}$, $V_{ph2}$ is line and phase voltage on secondary side

$$V_{ph1} = \frac{V_{L1}}{\sqrt{3}} \tag{2.26}$$

$$V_{ph2} = KV_{ph1} = K \times \frac{V_{L1}}{\sqrt{3}}$$

Since secondary is connected in delta

$$V_{ph2} = V_{L2}$$

$$V_{L2} = K \times \frac{V_{L1}}{\sqrt{3}} = \frac{K}{\sqrt{3}} \times V_{L1} \tag{2.27}$$

iv. Delta - Star (Δ – Y) connection

Figure 2.31 shows the primary winding connected in delta and secondary winding is connected in star connection.

$V_{L1}$, $V_{ph1}$ is line and phase voltage on primary side
$V_{L2}$, $V_{ph2}$ is line and phase voltage on secondary side
As primary winding is delta connected,

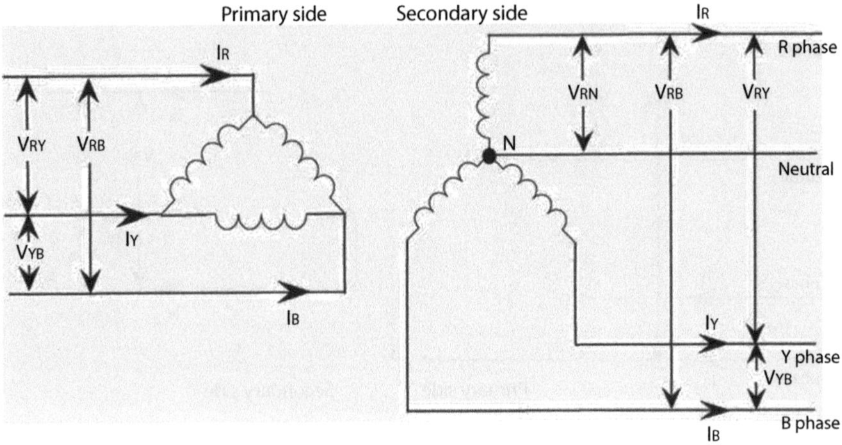


**Figure 2.31** Delta – Star configuration of three-phase transformer.

$$V_{L1} = V_{ph1} \tag{2.28}$$

$$V_{ph2} = KV_{ph1}$$

As secondary is star connected,

$$V_{L2} = \sqrt{3}V_{ph2}$$

$$V_{L2} = (\sqrt{3} \times K)V_{ph1}$$

$$V_{L2} = (\sqrt{3} \times K)V_{L1} \tag{2.29}$$

**C. Auto Transformer**

An auto transformer is one type of transformer, having only one winding, unlike the two windings in a normal transformer, and this winding is common to both primary and secondary winding in an auto transformer. A two-winding transformer provides an electrical isolation between the circuit while an auto transformer provides the isolation in the circuit [2, 3]. The voltage is stepped up or stepped down using an auto transformer. The reason for considering the auto transformer in place of a two-winding transformer is cost reduction in one copper winding.

Only one winding is wound on a laminated magnetic core. The single winding of the auto transformer is used as common to both primary and secondary. The auto transformer used as step-down purpose is shown in Figure 2.32.

The point 'AB' forms the primary winding having $N_1$ turns while 'BC' forms the secondary winding with $N_2$ turns for step-down auto transformer.

The auto transformer used as step-up purpose is shown in Figure 2.33.

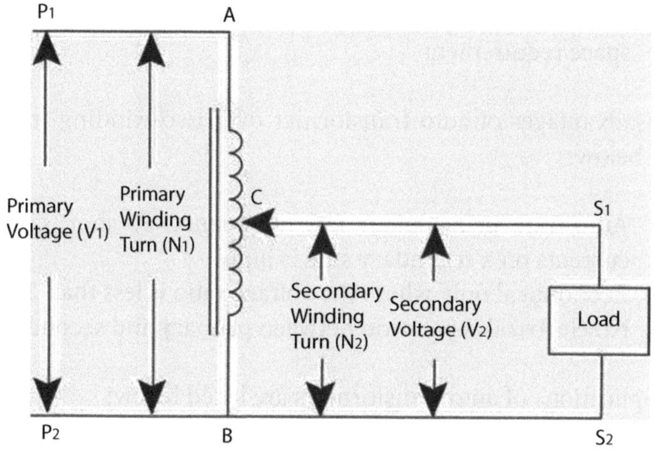


**Figure 2.32** Winding configuration step-down auto transformer.

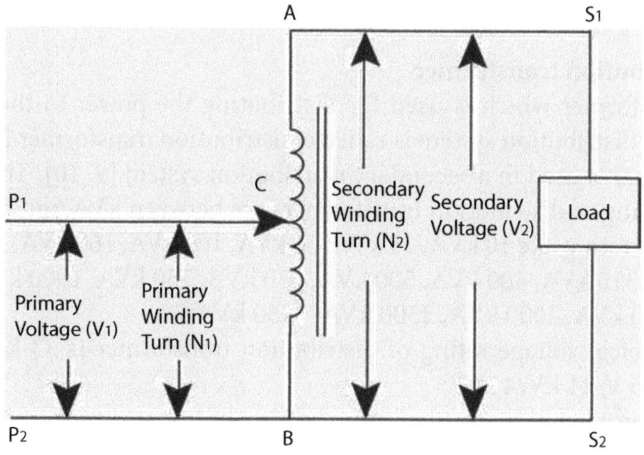


**Figure 2.33** Winding configuration step-up auto transformer.

The point 'AB' forms the secondary winding having $N_2$ turns while 'BC' forms the primary winding with $N_1$ turns for step-up auto transformer.

The advantages of auto transformer over two-winding transformer are listed below:

i. Copper requirement is less
ii. The efficiency is high
iii. The resistance and leakage reactance is less
iv. The copper loss ($I^2R$) is low
v. The cost is less
vi. The size of the auto transformer is less so it requires lesser space requirement

The disadvantages of auto transformer over two-winding transformer are listed below:

i. Auto transformer offers low impedance. So short circuit currents on a secondary side is high
ii. Economical only where the voltage ratio is less than 2
iii. No electrical separation between primary and secondary

The applications of auto transformers are listed below:

i. For starting of AC machines as a starter
ii. As a furnace transformer to supply power to the furnace at the required supply voltage
iii. For interconnecting the power system

**D. Distribution transformer**

The transformer which is used for distributing the power to the various loads in a distribution system is called a distribution transformer [12]. This transformer is used in a secondary distribution system [9, 10]. The typical power rating of distribution transformers are between kVA and few MVA in various rating like 10 kVA, 25 kVA, 50 kVA, 100 kVA, 160 kVA, 200 kVA, 250 kVA, 315 kVA, 400 kVA, 500 kVA, 630 kVA, 750 kVA, 1000 kVA, 1250 kVA, 1500 kVA, 2000 kVA, 2500 kVA, 3150 kVA, etc.

The typical voltage rating of distribution transformer is 33 kV/415 V, 22 kV/415 V, 11 kV/415 V.

The typical winding configuration of distribution transformer is Delta – Star, Star – Star, Star – Delta.

The distribution transformers are available in oil type (ONAN) and dry type (AN) cooling [11].

The typical oil type distribution transformer is shown in Figure 2.34.

The typical dry type distribution transformer is shown in Figure 2.35.

**E. Power transformer**

The transformer which is used for transferring the high power at power station, primary and secondary transmission substation system is called power transformer [2, 3]. The typical rating of power transformers are in the range of a few MVA to several hundred MVA like 20 MVA, 25 MVA, 40 MVA, 50 MVA, 80 MVA, 100 MVA, 150 MVA, 200 MVA, 250 MVA, etc.

The typical voltage rating of power transformer is 11 kV/69 kV, 11 kV/110 kV, 11 kV/132 kV, 22 kV/69 kV, 33 kV/110 kV, 33 kV/132 kV, 33 kV/230 kV, 33 kV/400 kV, 400 kV/765 kV, etc.

The typical winding configuration of distribution transformer is Delta – Star, Star – Star, Star – Delta, Star - Star.

The power transformers are mostly available in oil type (ONAN, OFAN, OFAF).

The typical power transformer is shown in Figure 2.36.

**Figure 2.34** Typical oil type transformer.

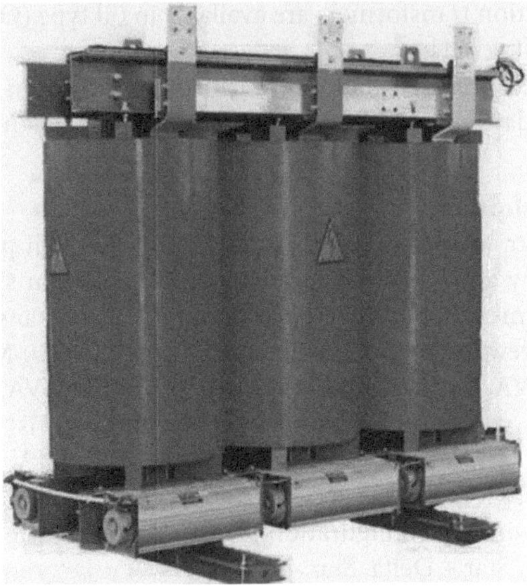

**Figure 2.35** Typical dry type transformer.

**Figure 2.36** Typical power transformer.

## 2.12 Components of the Transformer

The figure 2.37 shows the Silica gel used in a transformer.

**Figure 2.37** Silica gel.

The figure 2.38 shows the cooling tubes of the power transformer.

**Figure 2.38** Cooling tubes.

The figure 2.39 shows the low voltage bushing of the power transformer.

**Figure 2.39** Low voltage side bushing of power transformer.

The figure 2.40 shows the high voltage bushing of power transformer.

**Figure 2.40** High voltage side bushing of power transformer.

The figure 2.41 shows the low voltage side - star point earthing of power transformer.

**Figure 2.41** Low voltage side – star point earthing of power transformer.

The figure 2.42 shows the conservator tank of distribution transformer.

**Figure 2.42** Conservator tank of distribution transformer.

## 2.13 Standards for Transformers

The standards used for transformers are as follows:

1. International Electrotechnical Commission (IEC)
2. National Electrical Manufacturers Association
3. Bureau of Indian standards
4. Institute of Electrical and Electronics Engineers (IEEE)
   Etc.

### A. IEC standards for transformers

1. IEC 60076-1:2011 Power transformers - Part 1: General
2. IEC 60076-2:2011 Power transformers - Part 2: Temperature rise for liquid-immersed transformers
3. IEC 60076-3:2013+AMD1:2018 Consolidated version Power transformers - Part 3: Insulation levels, dielectric tests and external clearances in air
4. IEC 60076-4:2002 Power transformers - Part 4: Guide to the lightning impulse and switching impulse testing - Power transformers and reactors
5. IEC 60076-5:2006 Power transformers - Part 5: Ability to withstand short circuit
6. IEC 60076-6:2007 Power transformers - Part 6: Reactors
7. IEC 60076-7:2018 Power transformers - Part 7: Loading guide for mineral-oil-immersed power transformers
8. IEC 60076-8:1997 Power transformers - Part 8: Application guide
9. IEC 60076-10:2016 Power transformers - Part 10: Determination of sound levels
10. IEC 60076-10-1:2016 Power transformers - Part 10-1: Determination of sound levels - Application guide
11. IEC 60076-11:2018 Power transformers - Part 11: Dry-type transformers
12. IEC 60076-12:2008 Power transformers - Part 12: Loading guide for dry-type power transformers
13. IEC 60076-13:2006 Power transformers - Part 13: Self-protected liquid-filled transformers
14. IEC 60076-14:2013 Power transformers - Part 14: Liquid-immersed power transformers using high-temperature insulation materials

15. IEC 60076-15:2015 Power transformers - Part 15: Gas-filled power transformers
16. IEC/IEEE 60076-16:2018 Power transformers - Part 16: Transformers for wind turbine applications
17. IEC 60076-18:2012 Power transformers - Part 18: Measurement of frequency response
18. EC TS 60076-19:2013 Power transformers - Part 19: Rules for the determination of uncertainties in the measurement of the losses on power transformers and reactors
19. IEC TS 60076-20:2017 Power transformers - Part 20: Energy efficiency
20. IEC 60076-21:2018 Power transformers - Part 21: Standard requirements, terminology, and test code for step-voltage regulators
21. IEC 60076-22-1:2019 Power transformers - Part 22-1: Power transformer and reactor fittings - Protective devices
22. IEC 60076-22-2:2019 Power transformers - Part 22-2: Power transformer and reactor fittings - Removable radiators
23. IEC 60076-22-3:2019 Power transformers - Part 22-3: Power transformer and reactor fittings - Insulating liquid to air heat exchangers
24. IEC 60076-22-4:2019 Power transformers - Part 22-4: Power transformer and reactor fittings - Insulating liquid to water heat exchangers
25. IEC TS 60076-23:2018 Power transformers - Part 23: DC magnetic bias suppression devices
26. IEC TR 60076-26:2020 Power transformers - Part 26: Functional requirements of insulating liquids for use in power transformers
27. IEC/IEEE 60076-57-1202:2017 Power transformers - Part 57-1202: Liquid immersed phase-shifting transformers
28. IEC/IEEE 60076-57-129:2017 Power transformers - Part 57-129: Transformers for HVDC applications
29. IEC 61378-1:2011 Converter transformers - Part 1: Transformers for industrial applications
30. IEC 61378-3:2015 Converter transformers - Part 3: Application guide
31. IEC 62032:2012 Guide for the Application, Specification and Testing of Phase-Shifting Transformers

    Etc.

**B. IEEE standards for transformers**

Some of the IEEE standards for transformers are as follows:

1. C57.12.10-2017 – IEEE Standard Requirements for Liquis-Immersed Power Transformer
2. C57.12.36-2017 – IEEE Standard Requirements for Liquid-Immersed Distribution Substation Transformers
3. C57.12.39-2017 – IEEE Standard for Requirements for Distribution Transformer Tank Pressure Coordination
4. IEEE C57.110 – Recommended Practice for Establishing Liquid Immersed and Dry-Type Power and Distribution Tr ansformer Capability when Supplying Non sinusoidal Load Currents
5. IEEE C57.19.01 – Performance Characteristics and Dimensions for Power Transformer and Reactor Bushings

**C. Indian standards for transformers**

Some of the IEEE standards for transformers are as follows:

1. IS 2026 Specification for Power Transformers
2. IS 1180 Outdoor distribution Transformer up to and including 100 kVA
3. IS 335 Specification for Transformer Oil
4. IS 2099 Specification for high voltage porcelain bushing
5. IS 649 Testing for steel sheets and strips and magnetic circuits
6. IS 4257 Dimensions for clamping arrangements for bushings
7. IS 7421 Specification for Low Voltage bushings
8. IS 3347 Specification for Outdoor Bushings
9. IS 6600 Guide for loading of oil Immersed Transformers
10. IS 2362 Determination of water content in oil for porcelain bushing of transformer
11. IS 6103 Testing of specific resistance of electrical insulating liquids
12. IS 10028 Installation and maintenance of transformers

## Interview questions and answers

1. What happens if a 50 Hz frequency rated transformer operated at 60 Hz frequency?

Answer:

Core: Induced EMF $E_{RMS} = 4.44FTB_m \times Ai$

When the operating frequency increases to 60 Hz from 50 Hz, the flux in the core is reduced to a magnitude of (50/60) = 0.83 Bm (Bm is the original flux density)

Hence the core is not utilized effectively. However, if the opposite happens, i.e., a 60Hz rated transformer operating at 50 Hz frequency, the flux in the core is increased beyond rated values and due to core saturation overheating can occur.

When the frequency increases, eddy losses on both core and structural parts increase with square of frequency. Eddy current loss = $K \times Bm^2 \times F^2$, this is true only if the flux density is unchanged, whereas here the flux density has reduced to 0.83 times the original value as a result of change in frequency. Hence, the effective eddy current loss remains theoretically unchanged. But practically, there will be a small rise in the overall core loss due to increased stray loss and small increase in hysteresis loss. Hysteresis loss = $K \times F \times B^{1.6}$ (K= **Hysteresis** constant)

Winding: due to increased skin effect, the effective winding resistance can increase the mild thermal aberrations.

Voltage regulation: due to higher leakage reactance and increased winding resistance (at 60Hz), the voltage regulation will become poor.

2. What is the purpose of bushings and insulating materials used for housing?

Answer:

This creates the conducting channel between the feeder and the windings. The bushing conductor carries the phase current from or to the transformer. Materials used for bushing housing are (class C solid insulations): Porcelain, Fiber Reinforced Polymer (FRP), silicone rubber and in rare cases epoxy is used. The typical bushing of low voltage side of the power transformer is shown below in Figure 2.43.

3. What is the purpose of breather?

Answer:

The purpose is to equalize the pressure inside the conservator with that of atmospheric pressure during oil expansion and contraction. During the process of breathing air in and out, moisture may seep in.

**Figure 2.43** Bushing of low voltage side of power transformer.

**Figure 2.44** Breather with silica gel.

In order to prevent moisture, silica gel is placed at the breather inlet. The typical breather with silica gel of the power transformer is shown below in Figure 2.44.

4. Why is there an oil cup in the breather?
Answer:

An oil cup is used to filter unwanted solid particles (such as smoke and dust) entering into the breather. While breathing in, the air flow path is via the partial opening, through the oil and into the bottom lid and vice versa while breathing out.

5. Is breather gas operated or electrically operated?
Answer:

The breather is gas operated and electrical power is not required for its operation.

6. Why is the oil in the expansion/conservator tank always filled to more than half of its capacity and less than its full capacity?
Answer:

For a given surface area of oil exposed to air (air inside the conservator tank), the oil volume can be maximized only if the oil volume is more than half the conservator volume. Which means, the ratio (R) of surface area to the oil volume is least when the conservator oil is more than half filled. We need to reduce the ratio R because, R indirectly reflects as the percentage dissolved gas in the transformer oil. When the ratio of air exposed surface area to the oil volume is less, the amount of dissolved gas per unit oil volume is reduced. When R is reduced, the rate of oil degradation is drastically reduced. The conservator tank is not fully filled to give sufficient space for oil expansion inside the tank.

7. What is the operating temperature of a transformer?
Answer:

The operating temperature means the insulation temperature of the winding.

Dry type transformers operate typically between 80°C - 150°C
Oil filled transformers operate in the range 55°C - 65°C
SF6 or gas insulated transformers operate between 40°C - 55°C

8. What is the range of a power transformer efficiency and why it is very high?
Answer:

The efficiency of the typical power transformers ranges from 98% to 99.8%. Unlike in the case of motors/generators, a transformer has no rotating parts, hence it can easily be designed for such high efficiencies.

9. Why is the transformer rated in kVA and not rated in kW?
Answer:

Let us assume a load draws a certain amount of power from the transformer. If the load power factor is poor, then the transformer has to supply the higher current despite the same real power (kW) supplied. Less than a certain value of power factor, for the same real power (kW) supplied, the current required to be supplied would become more than the rated winding current. Hence all the components need to be de-rated.

Hence, real power (kW) capacity cannot be a criterion for defining a transformer rating as kW.

10. Why is transformer neutral earthing preferred?
Answer:

Neutral earthing is preferred because it helps the relays to detect earth fault currents and protects the system.

11. Can Buchholz relay operate for overload/short circuit faults?
Answer:

No, it will only operate for internal faults. Whenever there is a minor internal fault in the transformer such as insulation fault between turns, breakdown of core of the transformer, core heating, the insulating transformer oil will be decomposed in different hydrocarbon gases, $CO_2$ and CO. The gases produced due to the decomposition of transformer insulating oil will accumulate in the upper part of the Buchholz container, which causes a fall of oil level in it.

12. Why is the tap changer compartment isolated?
Answer:

Tap changer operation is achieved by fast mechanical switching of the taps and this may often cause the arcing. Hence oil inside the tap changer compartment can degrade at a faster rate as compared with oil inside the transformer compartment. Hence, oil in these different compartments cannot mix and this would drastically reduce the dielectric strength of the inter-turn and inter-disk insulation, thus risking the life of the transformer. For this reason, the tap changer has a separate compartment and it has a dedicated cooling system.

13. Why are rectangular conductors widely used for transformer windings?
Answer:

Rectangular conductors result in better space utilization and better mechanical strength because each rectangular strand is completely laid over. In the case of circular conductors, the mechanical stress is concentrated in the contact point of two strands. This stress would be large enough to cause deformations in the winding during mal operation.

Ultimately, use of rectangular windings results in higher short circuit withstanding capacity of the transformer.

# References


1. IEC 60076-1:2011 Power transformers - Part 1: General.
2. D. P. Kothari and I. J. Nagrath, *Electric Machines*, 4th Edison, Tata McGraw Hill, New Delhi, 2010.
3. S. K. Sahdev, *Electric Machines*, Cambridge University Press, New Delhi, 2018.
4. Saurabh Kumar Mukerji, Ahmad Shahid Khan and Yatendra Pal Singh, *Electromagnetics for Electrical Machines*, CRC Press, Boca Raton, 2015.
5. B. L. Theraja and A. K. Theraja, *Electrical Technology*, volume 2, S. Chand publication, 2006.
6. Jacek F. Gieras, *Electrical Machines,* CRC press, Boca Raton, 2017.
7. IS 1180 Outdoor distribution Transformer up to and including 100 kVA.
8. IS 2026 Specification for Power Transformers.
9. P. Sivaraman, C. Sharmeela and D. P. Kothari, "Enhancing the voltage profile in distribution system with 40 GW of solar PV rooftop in Indian grid by 2022: a review" 1st International conference on Large scale grid integration renewable energy in India, September, 2017, New Delhi, India.
10. Turan Gonen, *Electric* Power *Distribution Engineering,* 3rd Edison, CRC Press, 2014.
11. IS 6600 Guide for loading of oil Immersed Transformers.
12. IEEE Std C57.12.36-2007 IEEE Standard Requirements for Liquid-Immersed Distribution Substation Transformers.
13. Sivaraman, P., and Sharmeela, C. (2020). Introduction to electric distribution system, Baseem Kahn, Hassan Haes Alhelou & Ghassan (Eds.), *Handbook of Research on New Solutions and Technologies in Electrical Distribution Networks,* (pp. 1–31). Hershey, PA: IGI Global.
14. Sivaraman, P., and Sharmeela, C. (2020). Existing issues associated with electric distribution system. In Baseem Kahn, Hassan Haes Alhelou & Ghassan (Eds.), *Handbook of Research on New Solutions and Technologies in Electrical Distribution Networks,* (pp. 1–31). Hershey, PA: IGI Global.

15. IEEE Std C57.12.80-2012 IEEE Standard Terminology for Power and Distribution Transformers.
16. Hemchandra Madhusudan Shertukde, *Distributed Photovoltaic Grid Transformers*, CRC Press, 2014.
17. Sivaraman, P., and Sharmeela, C. (2020). Power Quality and its Characteristics. In Sanjeevikumar Padmanaban, C. Sharmeela, Jens Bo Holm-Nielsen, *Power Quality in Modern Power Systems*, Elsevier.
18. Sivaraman, P., and Sharmeela, C. (2020). Power System Harmonics. In Sanjeevikumar Padmanaban, C. Sharmeela, Jens Bo Holm-Nielsen, *Power Quality in Modern Power Systems*, Elsevier.

# 3
# DC Machines

## 3.1 Introduction

Electrical machines are used to transfer the energy from one form to another form, i.e., transferring the energy from mechanical form to electrical form or electrical form to mechanical form. These energy conversions are called electromechanical energy conversion. The machines which deal with Direct Current (DC) are called DC machines [1, 2, 11]. Generally, DC electrical machines are classified into DC generators and DC motors.

### 3.1.1 DC Generators

A DC generator is a machine that converts given rotating mechanical input power into equivalent DC electrical power. A DC generator produces the power according to Faraday's laws of electromagnetic induction. As per this law, whenever the conductor rotates in a magnetic field it cuts the magnetic field due to this EMF induced on the conductor.

### 3.1.2 DC Motors

A DC motor is a machine that is used to convert the given DC electrical power into rotating mechanical power. It works on the principle of Lorentz Law, which states that "the current carrying conductor placed in a magnetic and electric field experience a force"; that force is called the Lorentz force.

### 3.1.3 Construction of DC Machines

The construction of the DC machine is shown in Figure 3.1.

P. Sivaraman, C. Sharmeela, A. Thaiyal Nayagi, and R. Mahendran. Basic Electrical and Instrumentation Engineering, (125–184) © 2021 Scrivener Publishing LLC

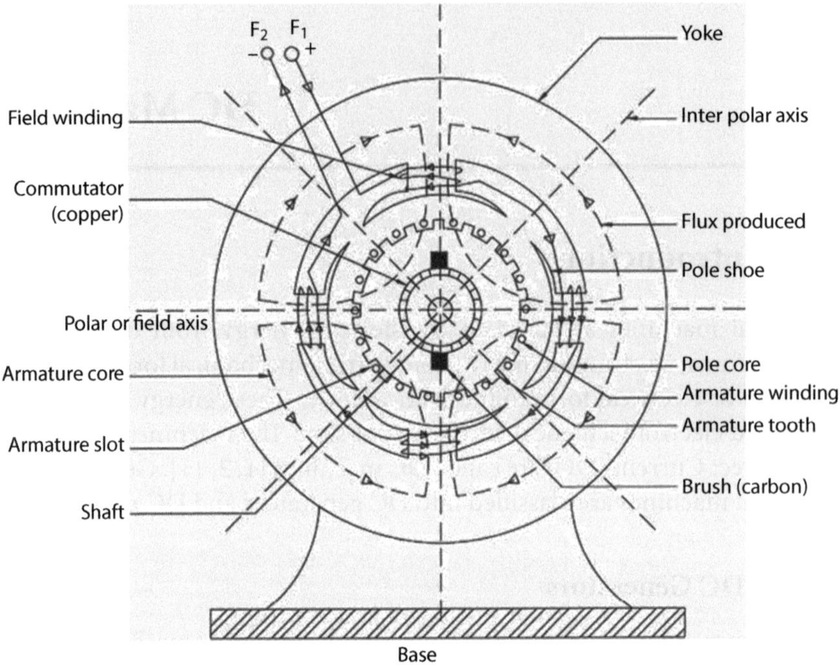


**Figure 3.1** Construction of the DC machine.

**A. Yoke**

Yoke is the outermost cover of the DC machine. It protects the equipments from dust, moisture and various gases likes acidic foams, $SO_2$, etc.

- Yoke provides the mechanical support to the pole
- Yoke provides the low reluctance path which provides low current and avoids wastage of power

In order to provide low reluctance path, rolled steel, cast steel, silicon steel is used.

**B. Poles**

Poles are part of the stator and are joined to the yoke with the help of bolts or welding. The pole carries the core and pole shoes are affixed on it. The pole shoes provide the two purposes

(i) Supporting field winding on it
(ii) Transfers the flux uniformly in air gap

**C. Field Winding**
Field winding coils are former wound and placed on each pole and are connected in series. It produces the magnetic field when exciting it by DC power supply.

**D. Armature**

**D1. Armature Core**
The typical armature of the DC machine is shown in Figure 3.2.

- It is the rotor of the DC machine
- It is made up of Cast iron or Cast Steel
- The shape of armature core is cylindrical and mounted on a shaft
- Air duct to permit the air flow through armature which serves cooling purpose
- It is made up of thin laminated circular steel disks to minimize the eddy current losses
- For cooling purpose, it may be provided with air ducts
- Armature core provides house for armature winding, i.e., armature conductor
- It will provide low reluctance path to the magnetic flux produced by the field winding

**Figure 3.2** Armature of the DC machine.

### D2. Armature Winding

The armature winding is the most important rotating part of the machine. It is the place where energy conversion takes place, i.e., the mechanical energy is converted into electrical energy (generator) and the electrical energy is converted into mechanical energy (motor).

- For generator, armature conductors cuts the magnetic flux gets and due to this EMF gets induced
- For motor, it will carry the current and produce the magnetic flux

The armature winding is mainly classified into types,

1. Lap winding
2. Wave winding

### i. Lap Winding

The typical lap winding is shown in Figure 3.3. The number of conductor parallel path of winding is equal to poles.

The lap winding is mainly used in low voltage and high current applications. They are classified into three types:

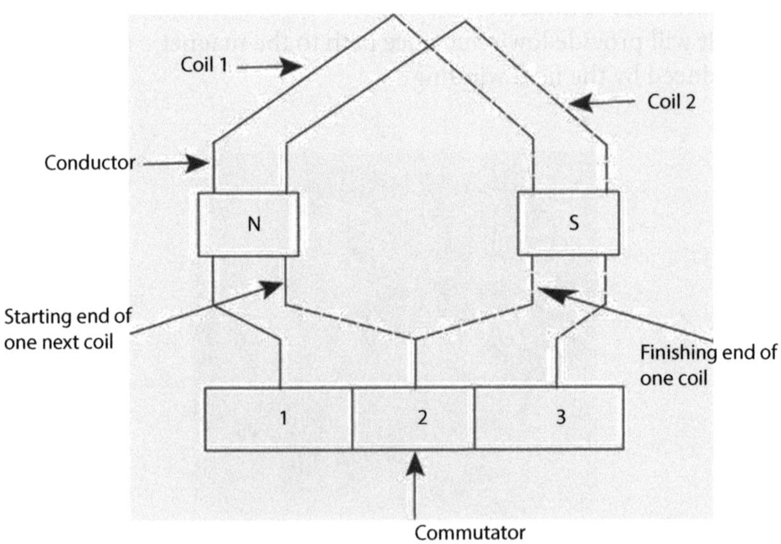


**Figure 3.3** Typical lap winding.

1. Simplex Lap Winding
2. Duplex Lap Winding
3. Triplex Lap Winding

**1. Simplex Lap Winding**

The one end of coil is connected to starting point of another coil as shown in Figure 3.4.

**2. Duplex Lap Winding**

The typical duplex lap winding is shown in Figure 3.5. This type of windings is used for high current applications. i.e. large motors.

**3. Triplex Lap Winding**

In triplex lap winding, the windings are connected to the one-third of the commutator bars.

The lap winding has many paths and hence it is used for the larger current applications. The only disadvantage of the lap winding is that it requires many conductors, which increases the winding cost.

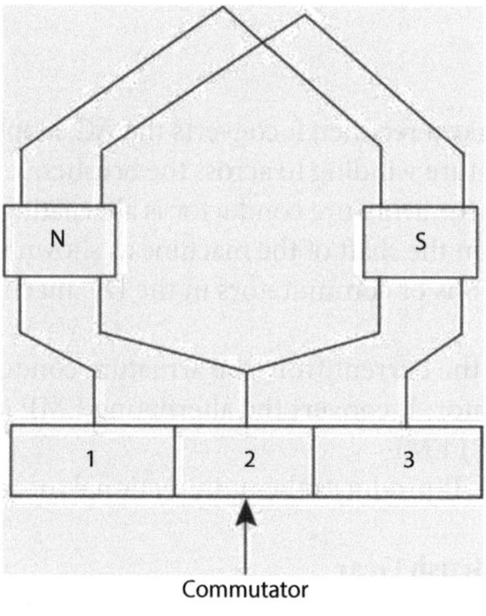


**Figure 3.4** Typical simple lap winding.

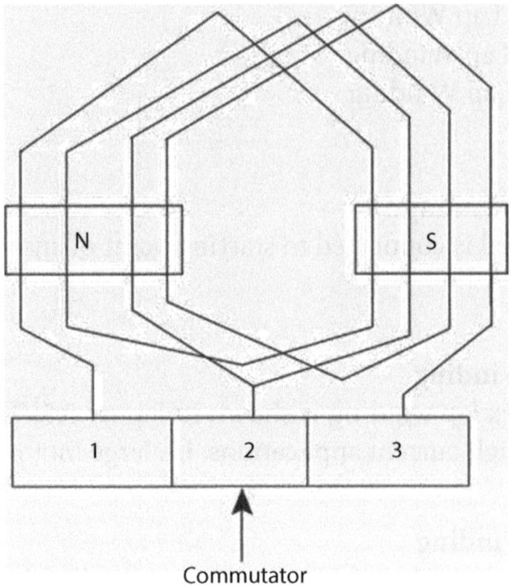


**Figure 3.5** Duplex lap winding.

### ii. Wave Winding

Where wave winding is employed, the number of parallel paths is equal to 2. i.e. two parallel paths are used between the brushes and winding, irrespective of poles. The typical wave winding is shown in Figure 3.6.

The difference between lap winding and wave winding are listed in Table 3.1.

### D3. Commutator

A Commutator is like a rectifier, it converts the AC supply into the DC supply within the armature winding to across the brushes. It is essential because of EMF induced in the armature conductor is alternating current. The commutator is located on the shaft of the machine as shown in Figure 3.7.

The main functions of commutators in the DC machines are as follows:

- It collects the current from the armature conductors
- For generator: It convert the alternating EMF into unidirectional (DC) EMF
- For motor: It produces the unidirectional torque

### D4. Brushes and Brush Gear

The electrical connection between the commutator and external circuit is achieved with the help of brushes. Generally it is made up of carbon.

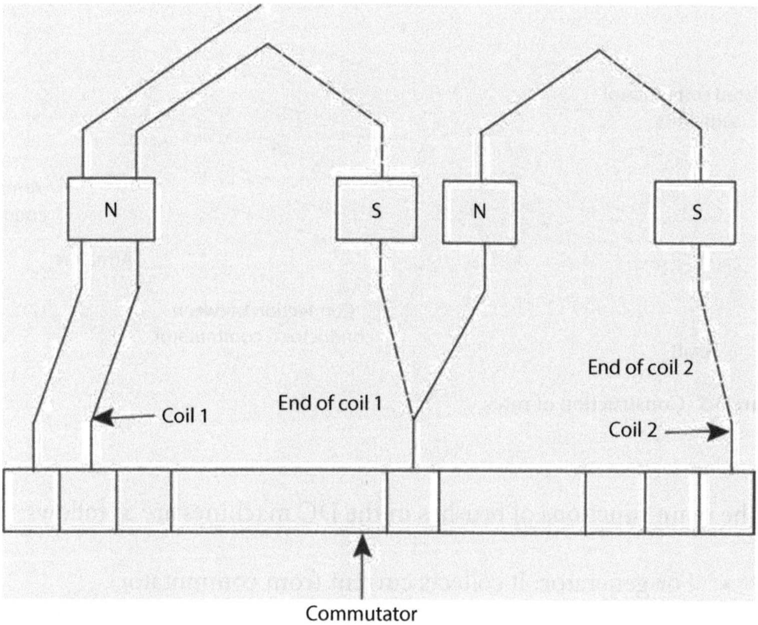


**Figure 3.6** Typical wave winding.

**Table 3.1** Difference between lap winding and wave winding.

| S. No. | Lap winding | Wave winding |
|---|---|---|
| 1 | The numbers of the parallel path are equal to the total number of poles | The number of the parallel path is equal to 2 |
| 2 | Another name of Lap winding is Parallel winding | Another name of Wave winding is Series winding |
| 3 | The EMF induced in this winding is less as compared with wave winding | The EMF induced in this winding is high as compared with lap winding |
| 4 | Efficiency is less as compared with wave winding | Efficiency is high as compared with lap winding |
| 5 | Cost of the lap winding is high because of larger number of coils as compared with wave winding | Cost of the wave winding is low because of fewer coils as compared with lap winding |
| 6 | It is used for high current and low voltage machines | It is used for low current and high voltage machine |

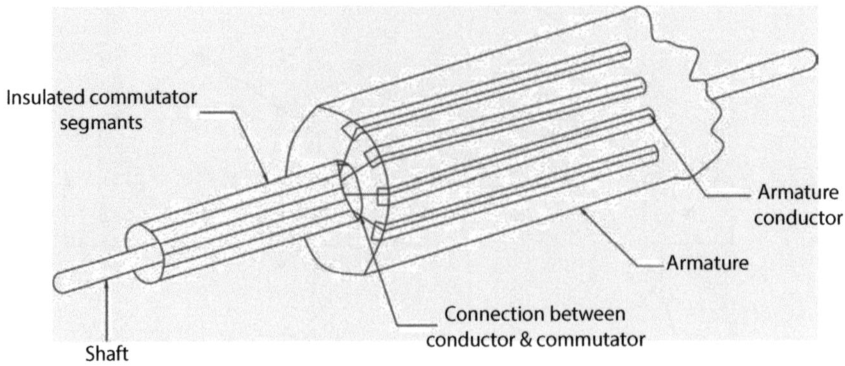


**Figure 3.7** Construction of rotor.

The main functions of brushes in the DC machines are as follows:

- For generator: It collects current from commutator
- For motor: It supplies the current from stationary external circuit to the armature winding through commutator

**D5. Bearings**

Ball bearings are usually used as they are more reliable. For heavy-duty machines, roller bearings are preferred.

## 3.2 Operation of DC Machines

### 3.2.1 Principle of DC Generators

The Fleming's right hand rule used to find the direction of induced EMF of the DC generator. i.e. Fleming's right hand rule used to determine the direction the current flows [1]. Whenever the conductor rotates in a magnetic field, an EMF is induced in the conductor as per Faraday's law of electromagnetic induction. Thumb, middle finger and index finger of right hand are kept perpendicular as shown in Figure 3.8,

- The Thumb shows the direction of the relative motion of the conductor with respect to the flux
- The Index Finger shows the direction of flux
- The Middle Finger shows the direction of the EMF induced in the conductor

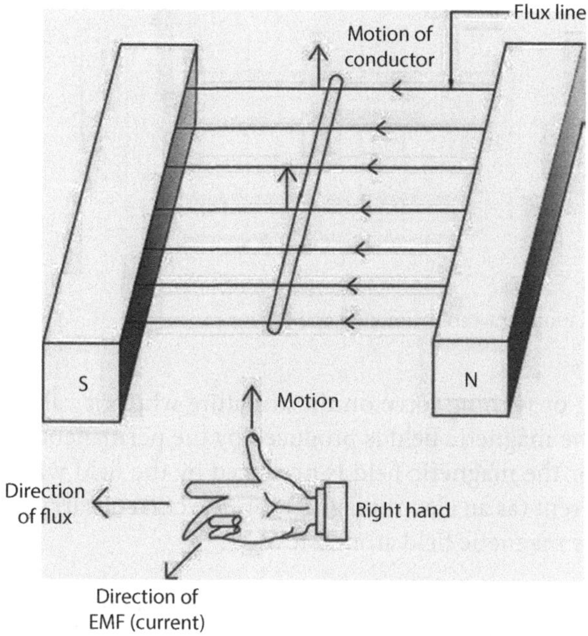


**Figure 3.8** Fleming right-hand rule.

The expression for induced EMF is given in equation 3.1.

$$E = B \times l \times V \tag{3.1}$$

Where

l is active length of conductor in m

V is relative velocity component of conductor in m/s in the direction perpendicular to direction of flux

The relation between the motion and flux is shown in Figure 3.9 (a) and (b).

The direction of the motion is perpendicular to the flux as shown in Figure 3.9 (a) -> induced EMF is maximum

The direction of the motion is parallel to the flux as shown in Figure 3.9 (b) -> induced EMF is zero

### 3.2.2 Operating Principle of Motors

When a current carrying conductor is placed in a magnetic field, it experiences a mechanical force. The force experienced by the conductor acts

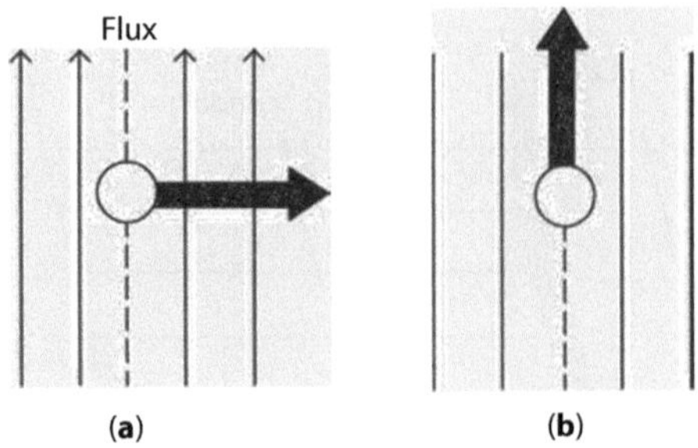


**Figure 3.9** Relation between the motion and flux.

as a twisting or turning force on the armature which is called torque [3]. Generally, the magnetic field is produced by the permanent magnet while in DC motor the magnetic field is produced by the field winding when it carries a current (as an electromagnet) [6]. The current carrying conductor produces the magnetic field around it.

- The flux produced by the permanent magnet called main flux as shown in Figure 3.10.
- The flux produced by the current carrying conductor as shown in Figure 3.11.

The conductor flux is opposed to the main flux as shown in Figure 3.12.

**LEFT:** Main flux and conduction flux are addition

**RIGHT:** Conductor flux and main flux are opposite, and therefore cancel each other

The direction of force is shown in Figure 3.13.

**Fleming Left-Hand Rule for DC Motor:**
Fleming's left-hand rule (for motor) shows the direction of motion in a motor windings. The left hand is held with the thumb, index finger and

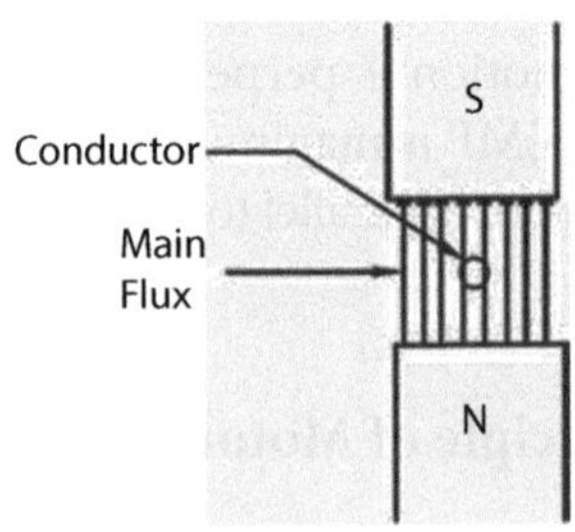


**Figure 3.10** Main flux by permanent magnet.

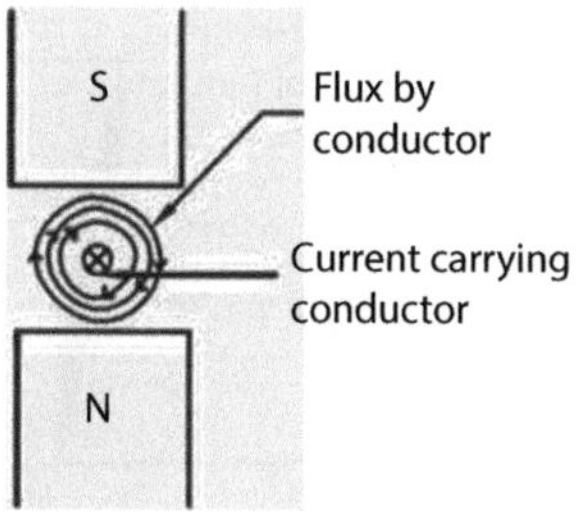


**Figure 3.11** Flux produced by current carrying conductor.

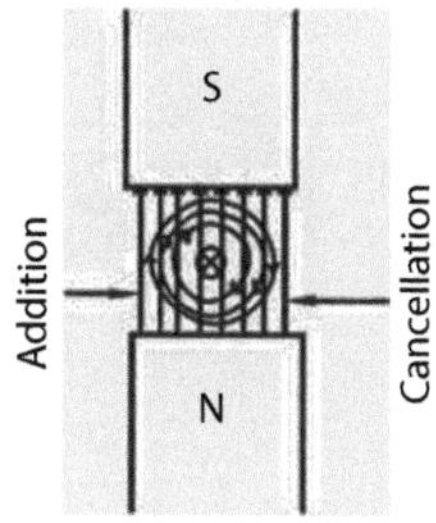


**Figure 3.12** Conductor flux is opposed to the main flux.

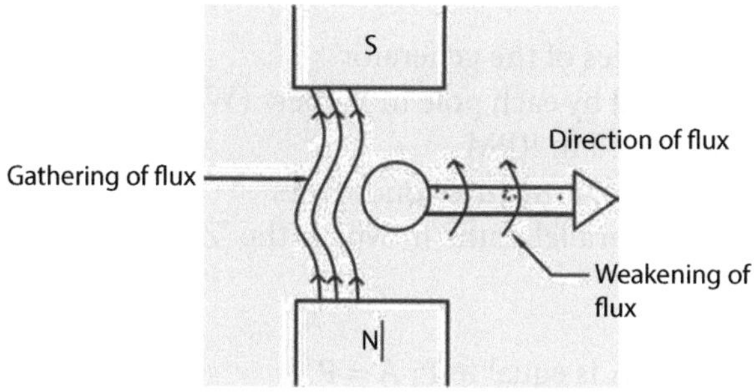


**Figure 3.13** Direction of force.

middle finger mutually perpendicular to each other (at right angles), as shown in Figure 3.14.

- The Thumb shows the direction of the motion
- The Forefinger shows the direction of the magnetic field
- The Middle finger shows the direction of the current.

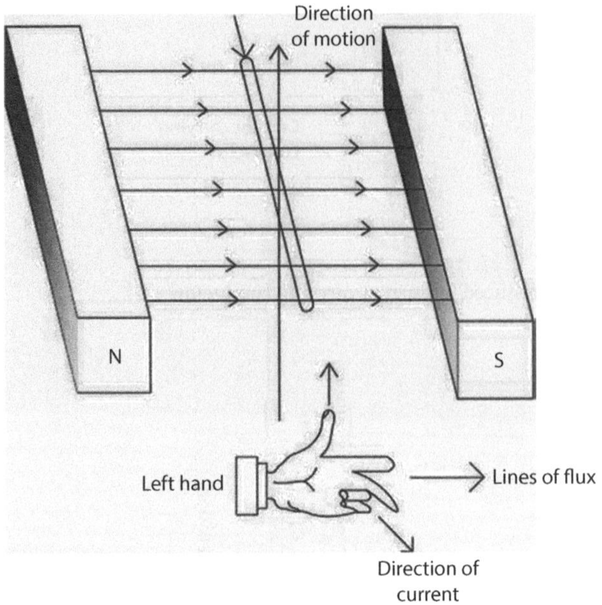


**Figure 3.14** Fleming left hand.

## 3.3 EMF Equation of DC Generator

Let

P is number of poles of the generator

Φ is flux produced by each pole in Webers (Wb)

N is armature speed in RPM

Z is total number of armature conductors

A is number of parallel paths in which the 'Z' numbers of conductors are divided

So

For lap winding, A is equal to P; A = P

For wave winding, A is equal to 2; A = 2

The EMF induced in the conductors are according to Faraday's Law of Electromagnetic Induction. Hence, the average value of EMF induced in each armature conductor is

$$\text{E} = \text{Rate of cutting the flux} = \frac{d\phi}{dt} \tag{3.2}$$

for considering one revolution, conductor cuts total flux produced by all the poles. i.e., $\phi \times P = \phi$. While time required to complete one revolution in 60/N seconds as speed (N) is in RPM.

$$E = \frac{\phi p}{60/N} = \frac{\phi PN}{60} \tag{3.3}$$

Conductors in one parallel path are always in series. There are total "Z" conductors with "A" parallel paths, hence (Z/A) number of conductors are always in series and EMF remains same across all the parallel paths.

Total induced EMF can be expressed as

$$E = \phi P \frac{N}{60} \times \frac{Z}{A} \text{ Volts} \tag{3.4}$$

This expression is called EMF equation of a DC generator.

For Lap Winding, A = P = $\frac{\phi PNZ}{60A}$ Volts

For Wave Winding A = 2 = $\frac{\phi PNZ}{120}$ Volts

**Example 3.1:**

A 4 Pole, Lap wound DC generator has a useful flux of 0.07 Wb per pole. Calculate the generated EMF when the armature speed at 900 RPM with the help of prime mover. Armature consists of 440 number of conductors. Also calculate the generated EMF if lap wound armature is replaced by wave wound armature.

**Solution:**

P = 4

Z = 440

$\phi$ = 0.07

N = 900 RPM

The total induced EMF in Volts from equation (3.4) is

$$E = \frac{\phi PNZ}{60A} \text{ Volts}$$

i) For Lap winding, A = P = 4

$$E=\frac{\phi NZ}{60}=\frac{0.07\times900\times440}{60}=462\ Volts$$

ii) For Wave winding A = 2

$$E=\frac{\phi PNZ}{120}=\frac{0.07\times\ 4\ x\ 900\times440}{120}=924\ Volts$$

## 3.4 Torque Equation of a DC Motor

The turning or twisting force required to twist the rotor about an axis is called torque [4]. Consider a wheel with radius R in meters acted upon by circumferential force "F" newtons as shown in Figure 3.15.

The wheel is rotating at the speed "N" in RPM then its angular speed is

$$\omega=\frac{2\pi N}{60}\ rad/_{sec} \tag{3.5}$$

So work done in one revolution is,

W = F * Distance travelled in one revolution = F × 2πR Joules

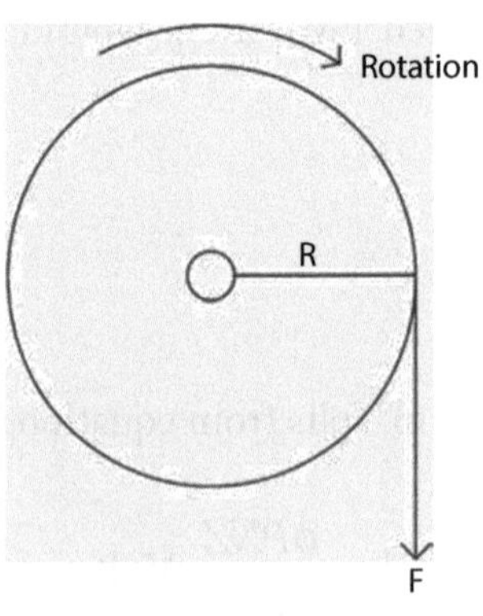


**Figure 3.15**

$$P = \text{Power developed} = \frac{\text{Work done}}{\text{Time}} = \frac{F \times 2\pi R}{\text{Time for one revolution}}$$

$$= \frac{F \times 2\pi R}{60/N} = (F \times R) \times \left(\frac{2\pi N}{60}\right)$$

$$P = T \times \omega \text{ in Watts}$$

Where,
T is torque in Nm
$\omega$ is angular speed in rad/sec

Let $T_a$ be the gross torque developed by the armature of the motor. It is also called armature torque.

Power in armature = Armature Torque × ω

i.e $E_b I_a = T_a \times \frac{2\pi N}{60}$

Back EMF ($E_b$) of a motor is given by,

$$E_b = \frac{\phi PNZ}{60A}$$

$$\therefore \frac{\phi PNZ}{60A} \times I_a = T_a \times \frac{2\pi N}{60}$$

$$\therefore T_a = \frac{1}{2\pi} \phi\, I_a \times \frac{PZ}{A} = 0.159\phi\, I_a \times \frac{PZ}{A}\, Nm \tag{3.6}$$

This equation is called as torque equation of a DC motor.

## 3.5 Circuit Model

The armature circuit has induced EMF ($E_a$) and it has the series resistance

$$R_a = \frac{R_p}{A'}$$

Where
$R_p$ is the resistance of a parallel path

The brush voltage drop is constant or the order of 2V, which is either ignored or its effect included in $R_a$ by linearizing it.

The field has a resistance $R_f$ fed with field current, $I_f = \frac{V_f}{R_f}$ and is shown at 90° to the brush axis as shown in Figure 3.16.

The field provides the per pole flux which induces the EMF ($E_a$) in the armature when it runs at speed (N).

### 3.5.1 Generator Mode

The armature current ($I_a$) flows in the direction or $E_a$ as shown in Figure 3.16.

For the armature circuit,

$$\text{Terminal voltage (V)} = E_a - I_aR_a; E_a > V$$

$$P_{mech}(\text{in net}) = E_aI_a(\text{mechanical power converted to electrical power})$$

$$P_{mech}\ (\text{in gross}) = E_aI_a + \text{rotational loss}$$

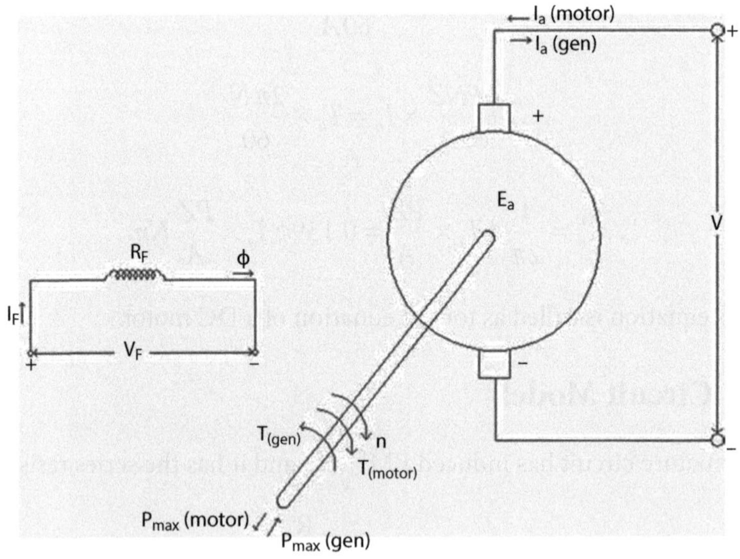


**Figure 3.16** Circuit model.

Where,

Rotational losses is equal to mechanical loss + core loss

$$P_o \text{ (electrical output)} = E_a I_a - I_a^2 R_a \text{ (armature copper loss)} \quad (3.7)$$

### 3.5.2 Motor Mode

The armature current ($I_a$) flows in the opposite direction to $E_a$ in a motor is called as back EMF.

For the armature circuit,

$$\text{Terminal voltage (V)} = E_a + I_a R_a; \; V > E_a$$

$$P_i \text{ (electrical input)} = VI_a$$

Where,

$$E_a I_a \text{ (electrical power converted to mechanical form)}$$

$$P_O = VI_a - I_a^2 R_a \text{ (armature copper loss)}$$

$$P_{mech} \text{ (out gross)} = E_a I_a$$

(or)

$$P_{mech} \text{ (out gross)} = E_a I_a - \text{rotational Losses} \quad (3.8)$$

### 3.5.3 Symbolic Representation of DC Generator

The armature is denoted by a circle with two brushes. Mechanically it is connected to another device called prime mover. The two ends of the armature are denoted as $A_1$ and $A_2$. The field winding is shown near the armature and the two ends are denoted as $F_1$ and $F_2$ as shown in Figure 3.17. The representation of field will vary based on the type of generator.

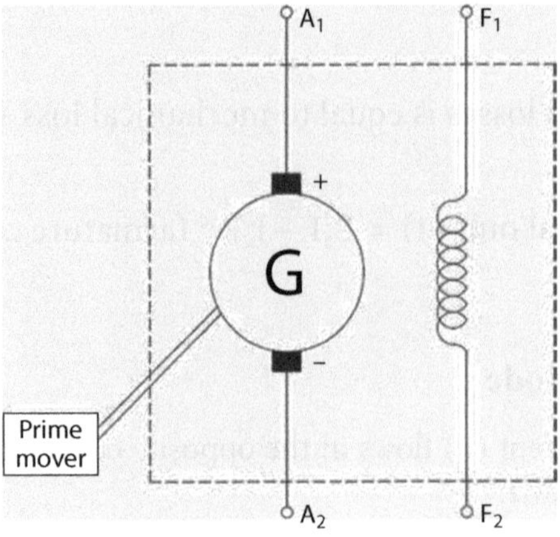


**Figure 3.17** Representation of DC generator.

## 3.6 Methods of Excitation

The magnetic field required for the operation of a DC generator is produced by an electromagnet. This electromagnet carries a field winding which produces the required magnetic flux when current is passed through it. The field winding is also called exciting winding, and current carried by the field winding is called exciting current [5, 10]. The supplying current to the field winding is called excitation and the way of supplying the exciting current is called method of excitation.

There are two methods of excitation used for DC Generators:

1. Separate excitation
2. Self-excitation

Depending upon the method of excitation method, the DC generators are classified into two types:

1. Separately excited generator
2. Self-excited generator

**A. Separately Excited DC Generator**

When the field winding is powered by the separate external DC supply (i.e., excitation of field winding is separate) then the generator is called separately excited generator. The circuit diagram of a separately excited DC generator is shown in Figure 3.18.

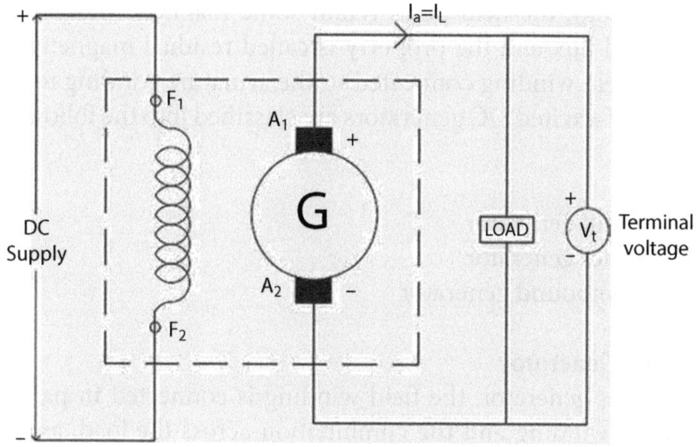


**Figure 3.18** Circuit diagram of separately excited DC generator.

Resistance of this field winding is high in order to limit the field current.

The field winding is excited separately, so the field current depends on supply voltage and resistance of the field winding.

The internally induced EMF (E) supplies the load, hence terminal voltage $V_t$ is a part of E and E is not equal to $V_t$ while supplying a Load.

The Voltage drop across the armature winding is equal to $I_aR_a$ Volts

$$E = V_t + I_aR_a + V_{brush} \tag{3.9}$$

Where,

$$I_a = I_L$$

$$\therefore E = \frac{\phi PNZ}{60A}$$

Where

$V_t$ is terminal Voltage

$I_aR_a$ is armature Resistance drop

$V_{brush}$ is brush contact drop (negligible)

**B. Self-excited DC Generator**

When the field winding of a generator is powered from the armature of the generator, then this generator is called a self-excited generator. When the generator is not working (i.e., without any current from armature winding to field winding), the field poles retain some magnetic flux. This flux is called residual flux and the property is called residual magnetism. Based on how the field winding connected to the armature winding to drive the excitation, self-excited DC generators are classified into the following three types:

i. Shunt generator
ii. Series generator
iii. Compound generator

**i. DC Shunt Generator**

In a DC shunt generator, the field winding is connected in parallel with the armature winding and the combination across the load, as shown in Figure 3.19.

The shunt generator field winding has a large number of turns of thin wire so it has high resistance ($R_{sh}$).

$I_a$ is armature current

$I_L$ is load current

$I_{sh}$ is shunt current

The current flow in armature winding ($I_a$) is the sum of load current ($I_L$) and field current ($I_{sh}$).

$$I_a = I_L + I_{sh}$$

Now voltage across load terminal is $V_t$ which is the same across field winding as both are in parallel with each other. The current flow through the field winding is given in equation 3.10.

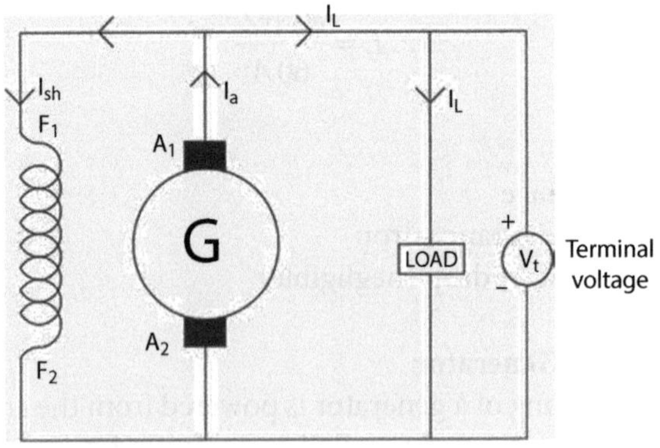


**Figure 3.19** DC shunt generator.

$$I_{sh} = \frac{V_t}{R_{sh}} \quad (3.10)$$

$$E = V_t + I_a R_a + V_{brush}$$

Where

$$E = \frac{\phi PNZ}{60A}$$

### ii. Series Generator

In a DC series generator, both the field winding and armature windings are connected in series. This generator is called a DC series generator. The resistance of series field winding is much less and it has a lower number of winding turns. The circuit diagram of DC series generator is shown in Figure 3.20.

$I_a$ is armature current
$I_L$ is load current
$I_{se}$ is series current

Current flow in the field winding is equal to armature winding as both windings are in series.

$$I_a = I_L = I_{se}$$

$$E = V_t + I_a R_a + I_a R_{se} + V_{brush}$$

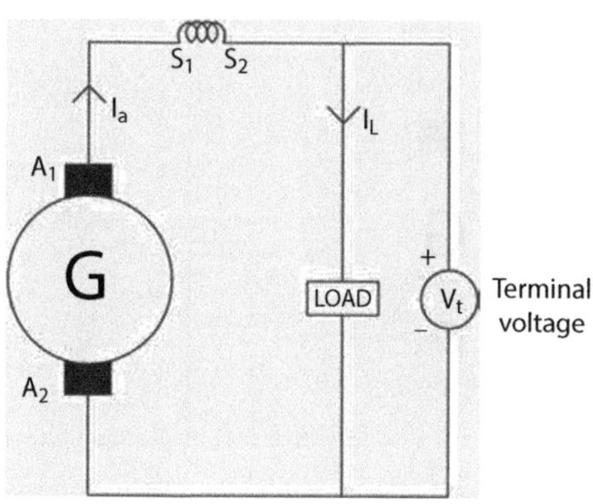


**Figure 3.20** Circuit diagram of DC series generator.

$$E = V_t + I_a (Ra + R_{se}) + V_{brush} \tag{3.11}$$

Where

$$E = \frac{\phi PNZ}{60A}$$

**iii. Compound Generator**

The DC compound generator is a combination of both DC series and DC shunt generator. The part of field winding is connected in parallel with the armature and part in series with the armature.

The DC compound generator is classified into the following two types:

a. Long shunt compound generator
b. Short shunt compound generator

**A. Long Shunt Compound Generator:**

In a long shunt compound generator, the shunt field winding is connected parallel with field winding and series with armature winding as shown in Figure 3.21.

$$I_a = I_{se} \text{ and}$$

$$I_a = I_{sh} + I_L$$

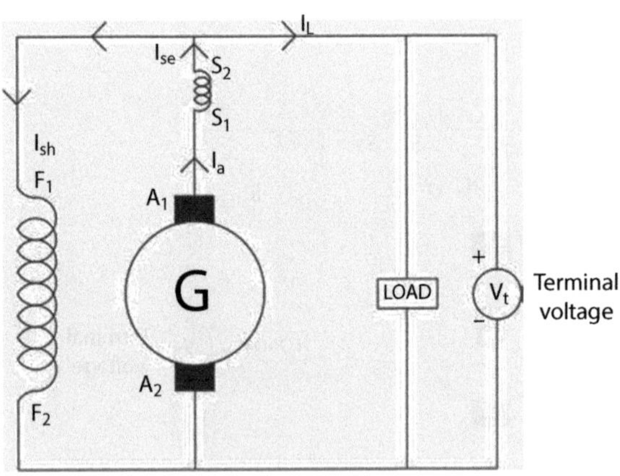


**Figure 3.21** Long shunt DC compound generator.

Voltage across shunt field winding is $V_t$ and current ($I_{sh}$) is

$$I_{sh} = \frac{V_t}{R_{sh}}$$

$$E = V_t + I_a R_a + I_a R_{se} + V_{brush} \tag{3.12}$$

Where
$R_{sh}$ is resistance of shunt field winding
$R_{se}$ is resistance of series field winding

**B. Short Shunt Compound Generator**

In a short shunt compound generator, only shunt field winding is connected in parallel with the armature winding excluding series field winding as shown in Figure 3.22.

The current flow in armature winding ($I_a$) is sum of the current flowing through series winding current ($I_{se}$) and shunt winding current ($I_{sh}$).

$$I_a = I_{se} + I_{sh}$$

Where,

$$I_{se} = I_L$$

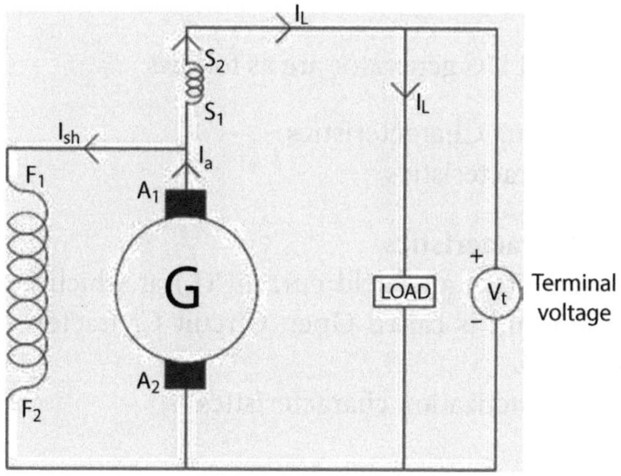


**Figure 3.22** Short shunt compound generator.

$$I_a = I_{sh} + I_L$$

The drop across shunt field winding is drop across the armature only and not the total voltage Vt, voltage drop across shunt field winding is E - $I_aR_a$.

$$I_{sh} = \frac{E - I_aR_a}{R_{sh}}$$

$$E = V_t + I_aR_a + I_{se}R_{se} + V_{brush}$$

$$I_{se} = I_L$$

$$E = V_t + I_aR_a + I_LR_{se} + V_{brush}$$

Neglecting $V_{brush}$,

$$E = V_t + I_aR_a + I_LR_{se}$$

$$E - I_aR_a = V_t + I_LR_{se}$$

$$I_{sh} = \frac{V_t + I_LR_{se}}{R_{sh}} \tag{3.13}$$

## 3.7 Characteristics of DC Generator

The characteristics of DC generator are as follows

i. Magnetizing Characteristics
ii. Load Characteristics

**i. Magnetizing Characteristics**

The no load voltage ($E_0$) and field current ($I_f$) at which generator speed is maintained constant is called Open Circuit Characteristics (OCC) as shown in Figure 3.23.

i.e. $E_O$VS $I_f$ is magnetization characteristics

Where

$E_O$ is no load induced EMF

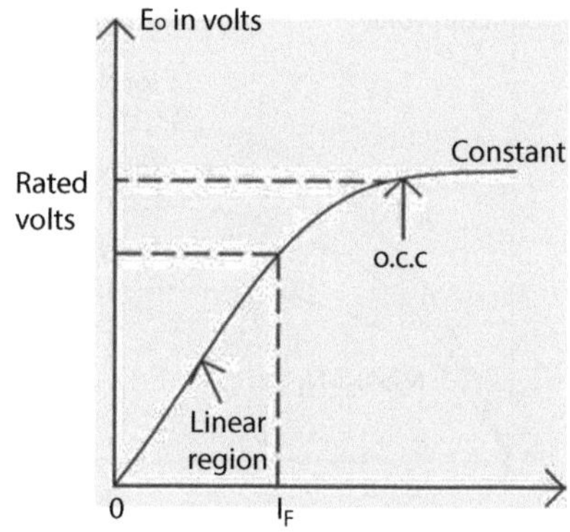


**Figure 3.23** No load characteristics at constant speed.

For generator,

$$E = \frac{\phi PNZ}{60A} \tag{3.14}$$

In equation 3.14, flux ($\phi$) and speed (N) are varying parameters.

a. Induced EMF by varying flux ($\phi$) and keeping speed (N) as constant

$$E \; \alpha \; \phi \text{ with } \frac{PNZ}{60A} \text{ constant}$$

$$E \; \alpha \; I_f \text{ as } \phi \; \alpha \; I_f$$

From Figure 3.23, induced EMF is directly proportional to field current ($I_f$). i.e. induced EMF increases with increase of field current until the saturation point. After certain value of $I_f$, the core gets saturated and flux $\phi$ also remains constant with increase of field current ($I_f$). Hence, after saturation the point, induced EMF remains constant.

b. Induced EMF by varying both flux ($\phi$) and speed (N)

$$E \, \alpha \, N \, \phi$$

For lower speeds, generated voltage is less. The no load characteristics for different speeds are shown in Figure 3.24.

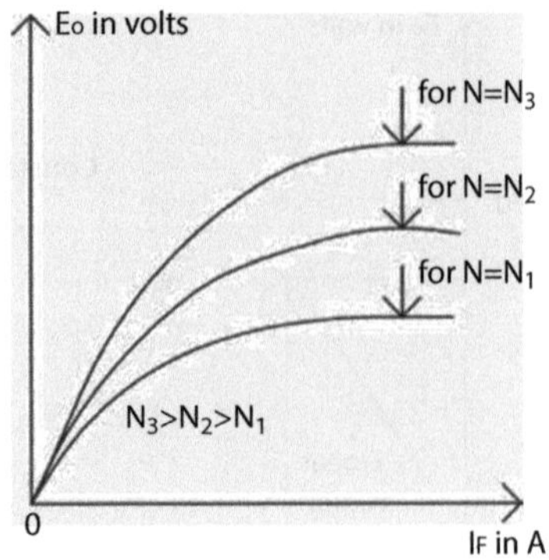


**Figure 3.24** No load characteristics for different speed.

## ii. Load Characteristics

a. External Characteristics: explains the terminal voltage ($V_l$) VS load current ($I_L$).
b. Internal characteristics: explains the generator induced EMF VS armature current ($I_a$).

### 3.7.1 Characteristics of Separately Excited DC Generator

1. Magnetization or open circuit characteristics
The circuit diagram of a separately excited DC generator is shown in Figure 3.25.

The rheostat is used as a potential divider to control the current flow to the field winding and the flux produced by the field winding. It varies from

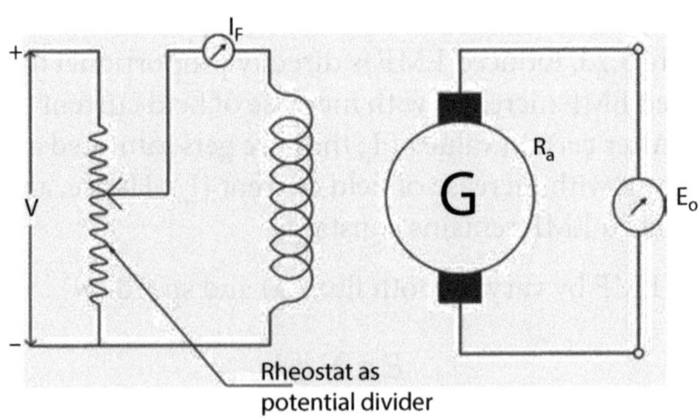


**Figure 3.25** Circuit diagram of separately excited DC generator.

zero to maximum value and current flow through the field is measured on the series connected ammeter. The EMF equation for DC generator,

$$E_O = \frac{\phi PNZ}{60A}$$

By varying the field current ($I_f$), flux ($\phi$) also changes. Hence, induced EMF ($E_0$) also varies. After the saturation point, flux becomes constant and hence induced EMF ($E_0$) become constant. The characteristics of a separately excited DC generator is shown in Figure 3.26.

2. Load curve

The terminal voltage ($V_t$) VS field current ($I_f$) of the separately excited DC generator is shown in Figure 3.27.

From Figure 3.27, the triangle PQR is called drop reaction triangle.

Where,
Region OP is no load saturation curve
Region OR is terminal voltage on load
Region OQ is generated voltage on load

3. Internal and External Characteristics

The induced EMF ($E_0$) be the rated no load voltage and its reduced to E due to armature reaction on no load condition. During the loading, further voltage is reduced to $V_t$ due to armature resistance drop $I_aR_a$ on load as shown in Figure 3.28.

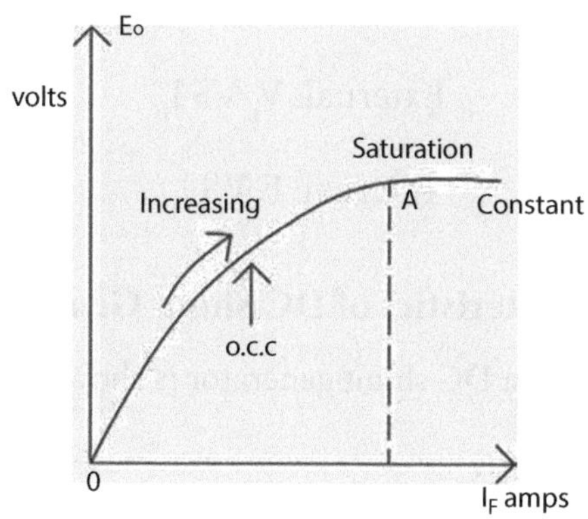


**Figure 3.26** Open circuit characteristics of separately excited DC generator.

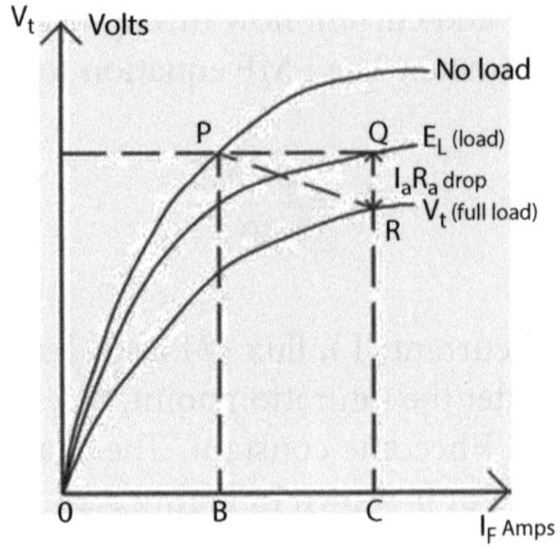


**Figure 3.27** Load characteristics of separately excited DC generator.

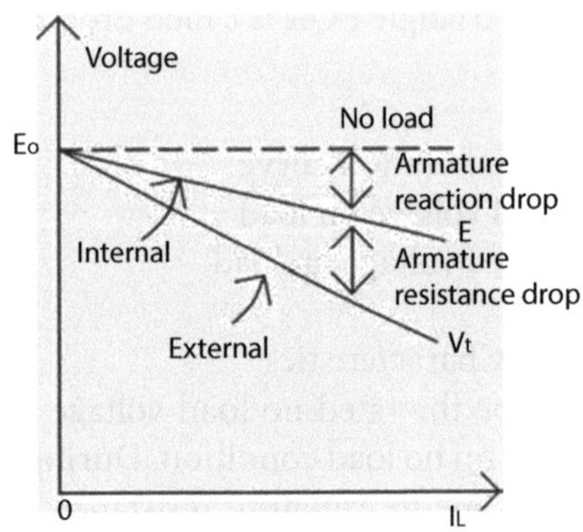


**Figure 3.28** Internal and external characteristics.

External: $V_t$ VS $I_L$

Internal: E VS $I_L$

### 3.7.2 Load Characteristics of DC Shunt Generator

The circuit diagram of a DC shunt generator is shown in Figure 3.29.

1. Internal Characteristics

The effect of flux produced by armature winding on the main flux produced by the field winding is called an armature reaction. Due to the armature reaction, the main flux pattern gets distorted. Hence, lesser flux gets

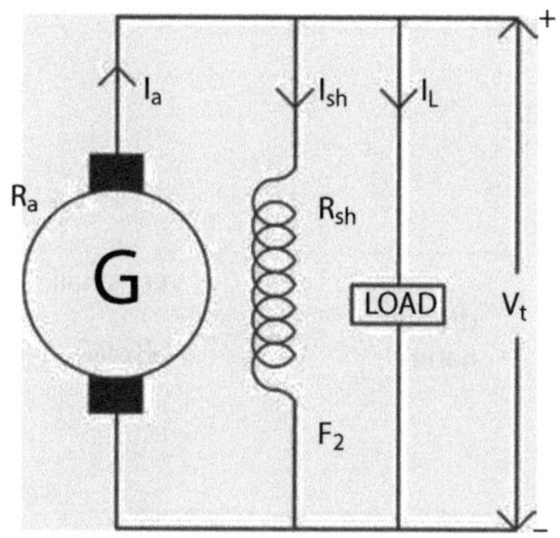


**Figure 3.29** Circuit diagram of DC shunt generator.

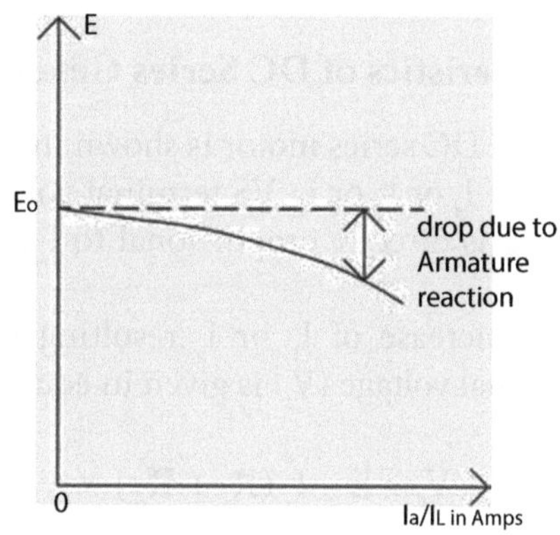


**Figure 3.30** Internal characteristics.

linked with the armature conductor and reduces the induced EMF. The characteristics are shown in Figure 3.30.

2. External Characteristics

The characteristics of terminal voltage ($V_t$) VS full load current ($I_L$) is shown in Figure 3.31.

The terminal voltage ($V_t$) is less than the generated voltage due to armature resistance drop. This can be written as,

$$V_t = E - I_a R_a \tag{3.15}$$

From equation 3.15, the increases of drop ($I_a R_a$) resulting to decrease of terminal voltage ($V_t$).

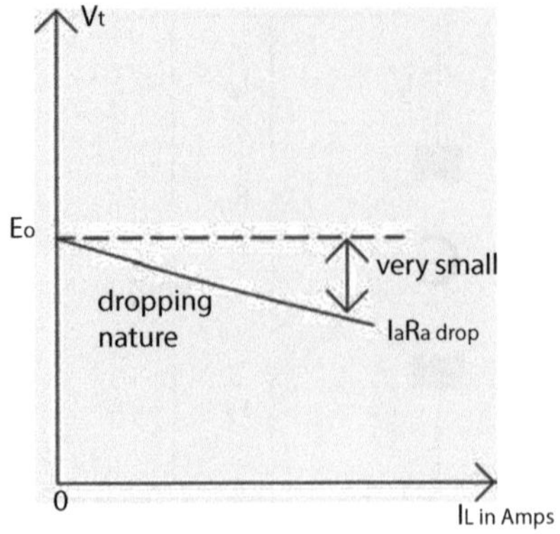


**Figure 3.31** External characteristics.

### 3.7.3 Load Characteristics of DC Series Generator

The circuit diagram of a DC series motor is shown in Figure 3.32.

The characteristics of $I_a$ or $I_L$ or $I_{se}$ VS terminal voltage ($V_t$) is shown in Figure 3.33. The flux $\phi$ is directly proportional to $I_{se}$, if flux increase and then E also increases.

From Figure 3.33, increase of $I_L$ or $I_a$ resulting to increase of drop $I_a(R_a+R_{se})$, hence terminal voltage ($V_t$) is given in equation 3.16.

$$V_t = E - I_a (R_a + R_{se}) \tag{3.16}$$

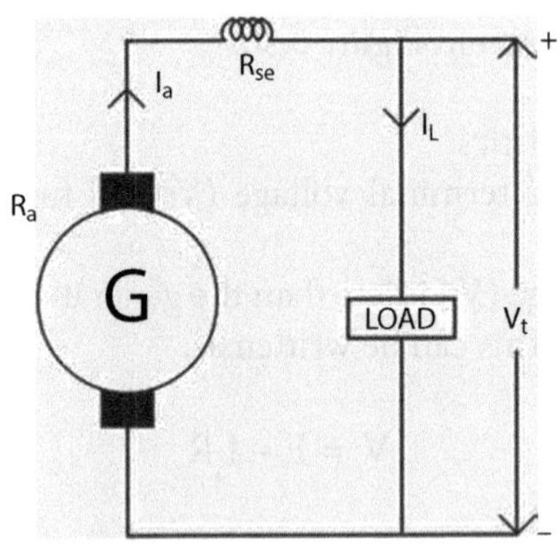


**Figure 3.32** Circuit diagram of DC series motor.

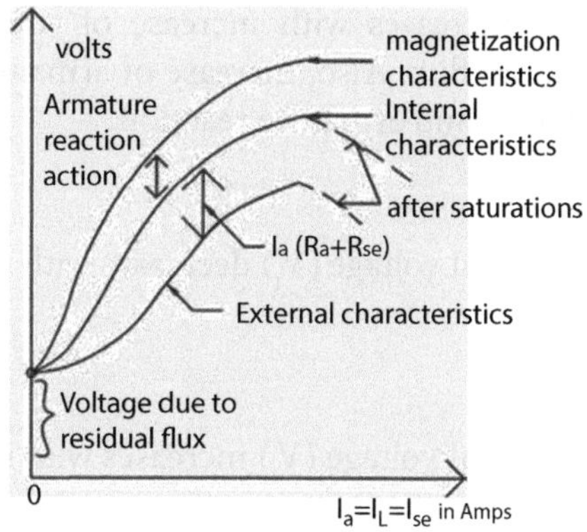


**Figure 3.33** Characteristics of $I_a$ or $I_L$ or $I_{se}$ VS terminal voltage ($V_t$).

In a DC series generator, open circuit characteristics cannot be obtained because $I_a = I_L = 0$. Hence induced EMF is zero.

### 3.7.4 Load Characteristics of DC Compound Generator

The full load current ($I_L$) VS terminal voltage ($V_t$) is shown in Figure 3.34.

1. Cumulatively Compounded

The total flux ($\phi_t$)is given in equation 3.17.

$$\phi_t = \phi_{sh} + \phi_{se} \tag{3.17}$$

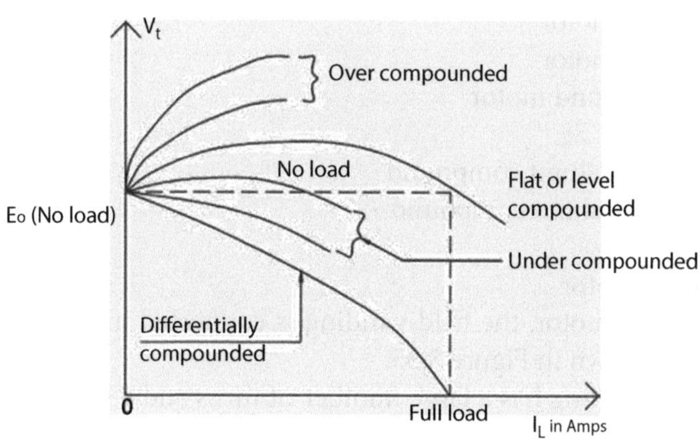


**Figure 3.34** Full load current VS terminal voltage ($V_t$).

The induced EMF increases with increase of armature current ($I_a$). Hence, it produces more flux. Also, increase of armature current ($I_a$) will increase the voltage drop and armature reaction.

2. Under Compounded
From Figure 3.34, terminal voltage ($V_t$) decreases with increase of full load current ($I_L$).

3. Over Compounded
From Figure 3.35, terminal voltage ($V_t$) increases with increase of full load current ($I_L$).

4. Flat or Level Compounded
The increase of load current ($I_L$), terminal voltage ($V_t$) remains constant, and it is called flat compounded.

5. Differential Compound
The load current ($I_L$) increases and flux ($\phi_{sh}$) is almost constant and $\varphi_{se}$ increases rapidly resulting in rapid decrease of terminal voltage ($V_t$).

## 3.8 Types of DC Motor

The DC motors are similar to DC generators in construction. These DC motors are classified into three types based on how the field winding is connected with the armature winding [7]. They are as follows:

1. Shunt motor
2. Series motor
3. Compound motor

   i. Short shunt compound
   ii. Long shunt compound

A. DC Shunt motor
In a DC shunt motor, the field winding is connected in parallel with the armature as shown in Figure 3.35.

The field winding has a large number of turns and a smaller cross-section area.

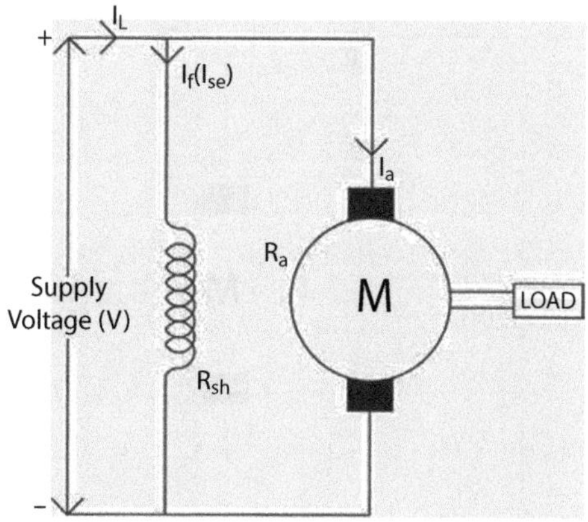


**Figure 3.35** Winding connection of DC shunt motor.

Where,
$R_a$ is resistance of the armature winding
$R_{sh}$ is resistance of the field winding

The supplied input voltage (V) to the armature winding and field winding is the same and load current ($I_L$) is equals to sum of field winding current ($I_{sh}$) and armature winding current ($I_a$). The expression for full load current of the DC shunt motor is given in 3.18.

$$I_L = I_a + I_{sh} \tag{3.18}$$

$$I_f = I_{sh} = \frac{V}{R_{sh}} \text{ and}$$

$$V = E_b + I_a R_a + V_{brush} \tag{3.19}$$

The flux ($\phi$) produced by the DC shunt motor is proportional to the field current ($I_{sh}$).

$$\phi \,\alpha\, I_{sh}$$

B. DC Series motor
In a DC series motor, both field winding and armature windings are connected in series to the supply voltage as shown in Figure 3.36. Let, $R_{se}$ is

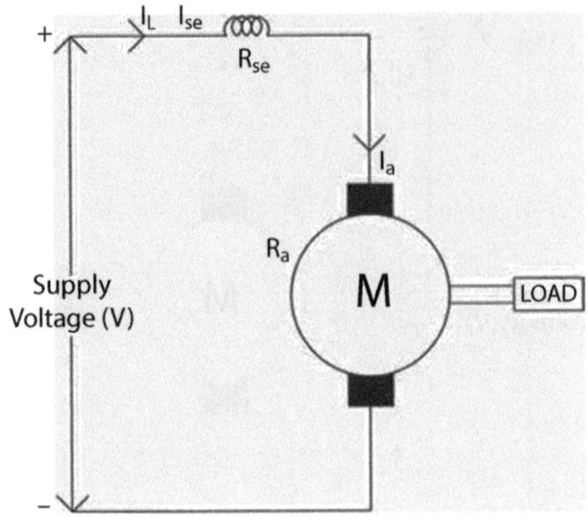


**Figure 3.36** Winding connection of DC series motor.

resistance of the series field winding and resistance of the winding is less; it is made of a lesser number of turns with a large cross-sectional area.

The full load current ($I_L$) is flowing through armature winding and field winding are the same.

$$I_L = I_{se} = I_a \tag{3.20}$$

$$V = E_b + I_a R_a + I_{se} R_{se} + V_{brush} \tag{3.21}$$

The flux ($\phi$) produced by the DC series motor is proportional to the field current ($I_{sh}$) or armature current ($I_a$).

$$\phi \; \alpha \; I_{se} \; \alpha \; I_a$$

C. DC Compound motor

The DC compound motor consists of part of the field winding connected in series and part of the field winding connected in parallel with armature winding. Based on the connections of field winding, a DC compound motor is classified into two types:

i. Long shunt compound motor
ii. Short shunt compound motor

i. Long Shunt Compound Motor

The shunt field winding is connected parallel to the combination of armature winding and the series field winding as shown in Figure 3.37.

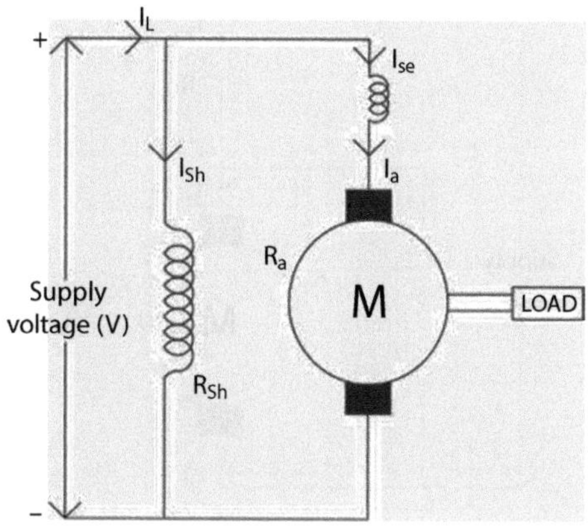


**Figure 3.37** Long shunt DC compound motor.

The full load current ($I_L$) is equal to the sum of current flowing through series winding ($I_{se}$) and current flowing through the shunt winding ($I_{sh}$). The expression for full load current ($I_L$) is given in equation (3.22).

$$I_L = I_{se} + I_{sh} \tag{3.22}$$

$$\therefore I_{se} = I_a$$

$$I_L = I_a + I_{sh}$$

$$I_{sh} = \frac{V}{R_{sh}}$$

$$V = E_b + I_a R_a + I_{se} R_{se} + V_{brush}$$

$$I_{se} = I_a$$

$$V = E_b + I_a (R_a + R_{se}) + V_{brush} \tag{3.23}$$

ii. Short Shunt Compound Motor

The shunt field winding connected in parallel to the armature winding only and the series field winding is connected in series with the combination as shown in Figure 3.38.

The full load current ($I_L$) is equal to current flow in the series winding ($I_{se}$) and given in equation (3.24).

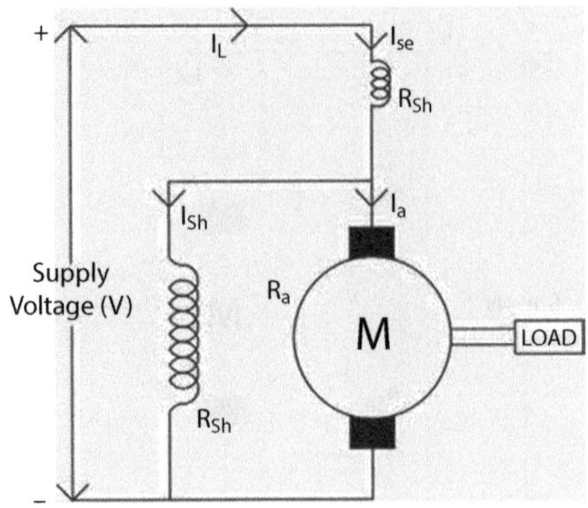


**Figure 3.38** Short shunt compound motor.

$$I_L = I_{se} \tag{3.24}$$

and

$$I_L \text{ or } I_{se} = I_a + I_{sh}$$

$$V = E_b + I_a R_a + I_{se} R_{se} + V_{brush}$$

$$I_{se} = I_L$$

$$V = E_b + I_a R_a + I_L R_{se} + V_{brush}$$

Drop across the shunt field winding is given in equation 3.25.

$$= V - I_L R_{se} = E_b + I_a R_a + V_{brush} \tag{3.25}$$

$$I_{sh} = \frac{V - I_L R_{se}}{R_{sh}} = \frac{E_b + I_a R_a + V_{brush}}{R_{sh}} \tag{3.26}$$

## 3.9 DC Motor Characteristics

The performance of a DC motor under various conditions can be obtained from the following characteristics:

1. Torque VS armature current (T VS $I_a$)
2. Speed VS armature current (N VS $I_a$)
3. Speed VS torque (N VS T)

### A. Characteristics of DC Shunt Motor

i. Torque VS. Armature Current Characteristics (T VS $I_a$)
The torque vs. armature current characteristics of a DC shunt motor is shown in Figure 3.39.

The torque developed by the DC shunt motor is directly proportional to armature current ($I_a$) and flux ($\phi$) as expressed in equation 3.27.

$$T \alpha \phi I_a \tag{3.27}$$

For a constant value of shunt winding resistance ($R_{sh}$) and supply voltage ($V_s$), current flow in the shunt winding ($I_{sh}$) is constant. Hence flux is also constant. Expression 3.27 becomes

$$T \alpha I_a \tag{3.28}$$

From equation (3.28), torque developed by the DC shunt motor linearly increases with the armature current. Whenever the load increases, current flow through the armature winding increases and torque developed by the motor linearly increases.

$T_{sh}$ (Shaft Torque) is less than the armature torque ($T_a$); the difference between these torque is called loss torque ($T_f$).

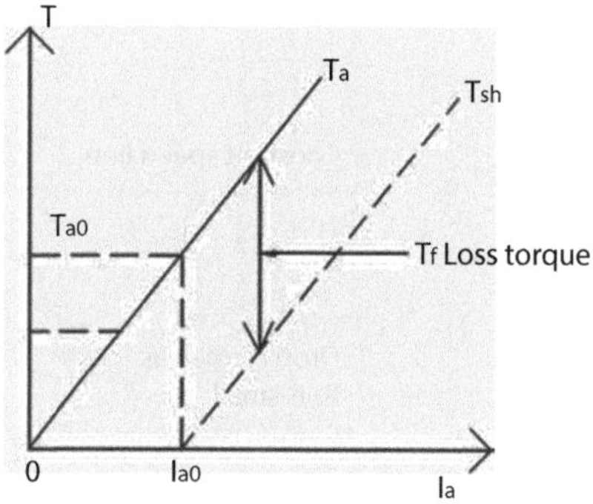


**Figure 3.39** Torque vs. armature current characteristics of a DC shunt motor.

ii. Speed vs. Armature current Characteristics (N VS $I_a$)

The speed vs. armature current characteristics of a DC shunt motor is shown in Figure 3.40.

From the speed equation,

$$N \alpha \frac{V - I_a R_a}{\phi} \tag{3.29}$$

$$N \alpha V - I_a R_a \text{ as } \phi \text{ is constant} \tag{3.30}$$

Whenever the load increases, current flow in the armature winding increases and hence drop $I_a R_a$ also increases. For constant supply voltage, $V - I_a R_a$ decreases and speed reduces. For small armature resistance ($R_a$), change in armature current ($I_a$) from no load to full load, drop $I_a R_a$ also very small and hence reduction in motor speed is not significant.

iii. Speed vs. Torque Characteristics (N VS T)

The speed vs. torque characteristics of a DC shunt motor is shown in Figure 3.41.

The speed is inversely proportional to torque of the motor.

**B. Characteristics of DC Series Motor**

i. Torque vs. Armature Current (T VS $I_a$)

The torque vs. armature current characteristics of a DC series motor is shown in Figure 3.42.

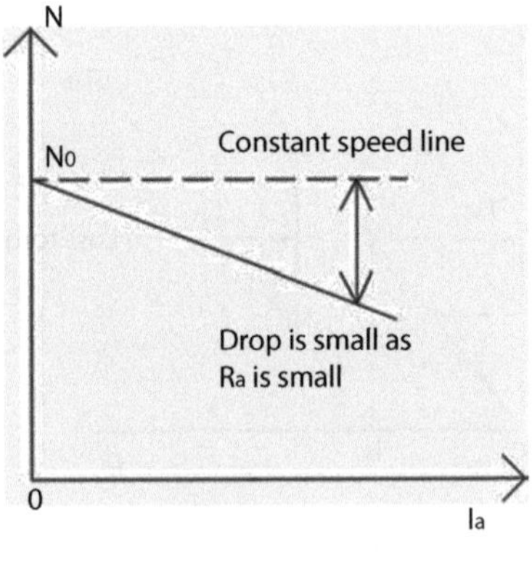


**Figure 3.40** Speed vs. armature current characteristics of a DC shunt motor.

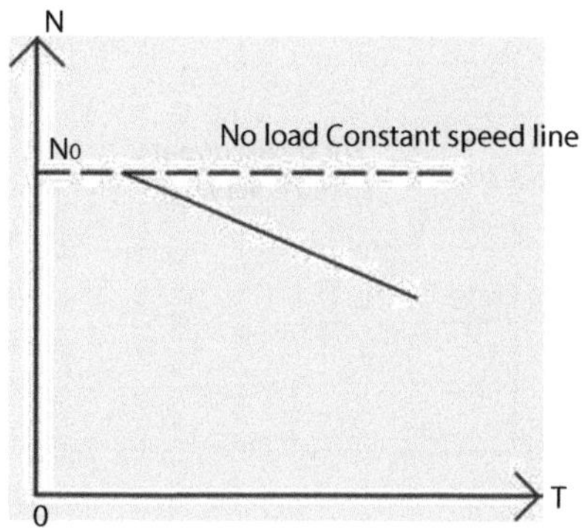


**Figure 3.41** Speed vs. torque characteristics of a DC shunt motor.

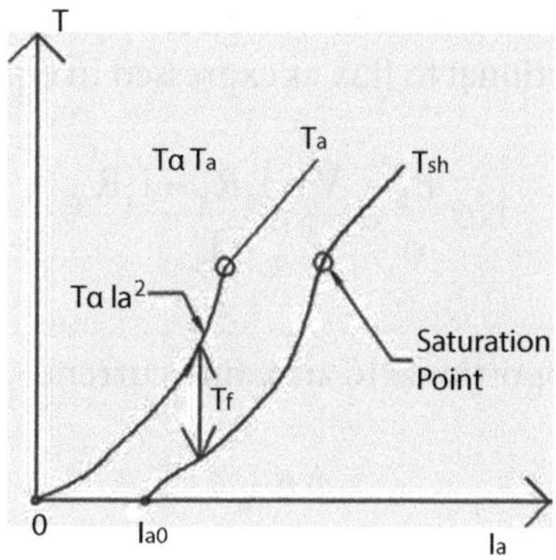


**Figure 3.42** Torque vs. armature current characteristics of a DC series motor.

The torque developed by the DC series motor is directly proportional to square of the armature current ($I_a$) as expressed in equation 3.31.

$$T \alpha I_a^2 \tag{3.31}$$

From equation (3.31), torque developed by the DC series motor linearly increase with the armature current. Whenever the load increases, current flow through the armature winding increases proportional to square of the armature current up to a certain limit. Due to saturation effect, flux produced by the winding remains constant with increase of current.

ii. Speed vs. Armature Current (N VS $I_a$)
The speed vs. armature current characteristics of a DC series motor is shown in Figure 3.43.

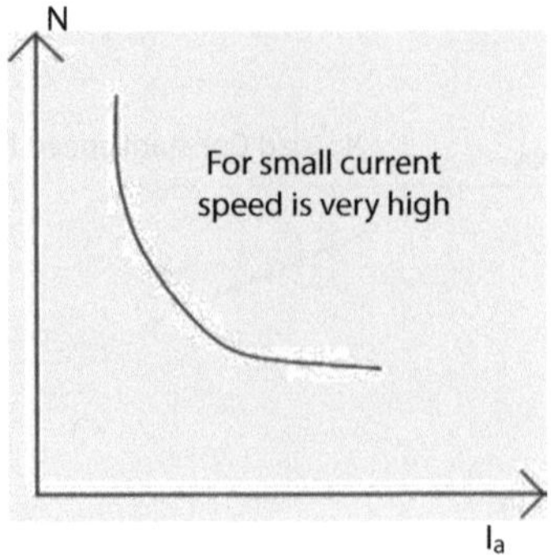


**Figure 3.43** Speed vs. armature current characteristics of a DC series motor.

The speed of the DC series motor is directly proportional to back EMF and indirectly proportional to flux as expressed in equation 3.32.

$$N\alpha\frac{E_b}{\phi}\alpha\frac{V-I_aR_a-I_aR_{se}}{I_a} \tag{3.32}$$

Flux is directly proportional to armature current $I_a$.

$$\phi\ \alpha\ I_a$$

The resistance value of $R_a$ and $R_{se}$ is small, so the effect of change in armature current ($I_a$) overrides the effect of change in $V - I_aR_a - I_aR_{se}$ on the speed. Hence, the speed regulation $E_b \cong V$ assumed as constant and speed reduces to $N\alpha\frac{1}{T_a}$.

iii. Speed vs. Torque Curve (N VS T)

The speed vs. torque characteristics of a DC series motor is shown in Figure 3.44.

The torque developed by the DC series motor is directly proportional to square of the armature current ($I_a$) as expressed in equation 3.31.

$$T\ \alpha\ I_a^2$$

$$N\ \alpha\ \frac{1}{I_a}$$

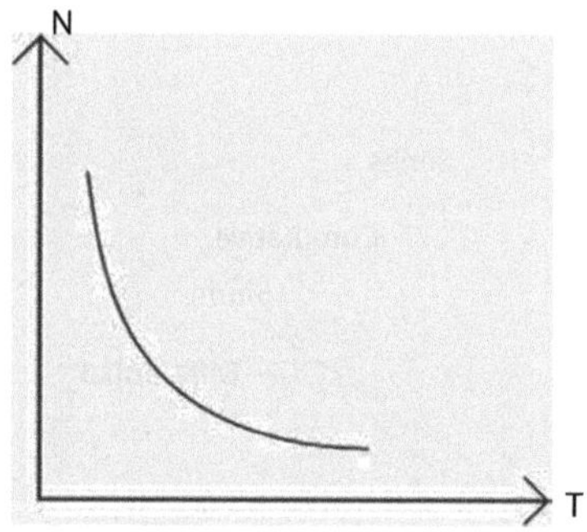


**Figure 3.44** Speed vs. torque characteristics of a DC series motor.

$$\therefore N \propto \frac{1}{\sqrt{T}} \tag{3.33}$$

From equation 3.33, speed increases with reduction of torque developed by the motor.

**C. Characteristics of DC Compound Motor**

The torque vs. armature current characteristics is shown in Figure 3.45, speed vs. armature current characteristics is shown in Figure 3.46. The speed vs. torque characteristics is shown in Figure 3.47.

The exact shape of these characteristics depends on the relative contribution of series and shunt field windings.

1. Cumulatives

The torque developed at low speed is sum of flux in shunt winding ($\phi_{sh}$) and flux in series winding ($\phi_{se}$).

2. Differential

The two flux oppose each other ($\phi_{sh}$ - $\phi_{se}$) and the resultant flux decreases as load increases.

## 3.10 Necessity for Starters

The DC motors are self-starting motors. Whenever the armature winding and the field winding receive the DC input power supply, motoring action takes place. During the starting time, DC motors draw the higher starting

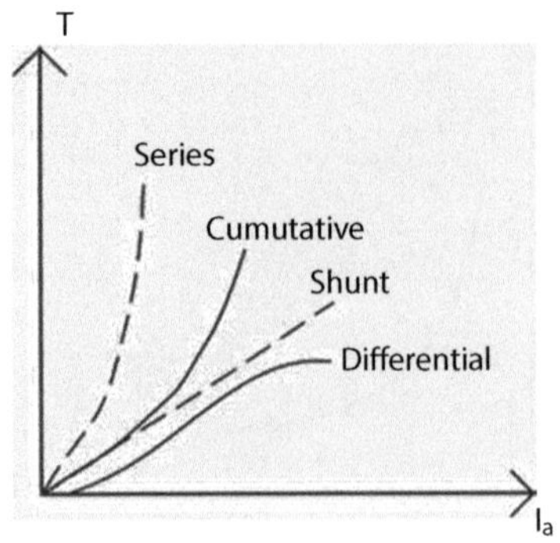


**Figure 3.45** Torque vs. armature current characteristics of a DC compound motor.

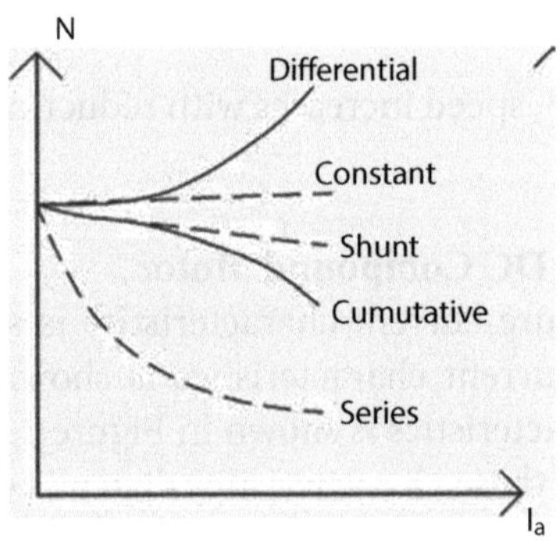


**Figure 3.46** Speed vs. armature current characteristics of a DC compound motor.

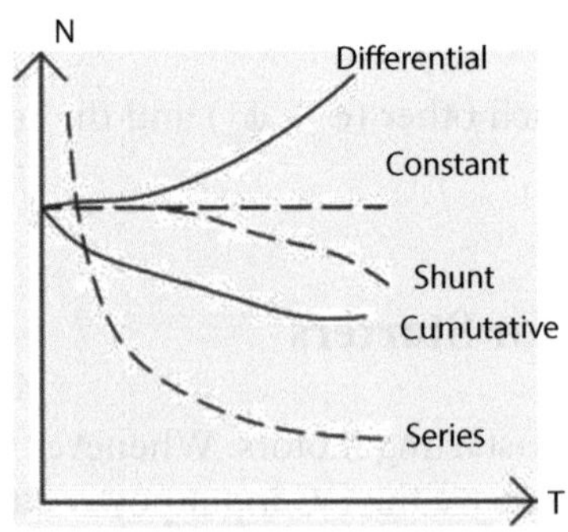


**Figure 3.47** Speed vs. torque characteristics of a DC compound motor.

current from the power supply. Hence, the starter is required to start a DC motor in a safe way [8]. The starter can limit the starting current drawn by the motor by adding the resistance across the power supply to its winding [9]. The expression for current limit by the starter during motor starting time is given in equation 3.34.

$$I_a = \frac{V}{R_a} \tag{3.34}$$

This resistance is called starter or a starting resistance. So basically a starter is a current limiting device during the starting time. There are three types of starters used to start the DC motors:

1. Two-point starter
2. Three-point starter
3. Four-point starter

**A. Three-Point Starter**

The circuit diagram of a three-point starter is shown in Figure 3.48.

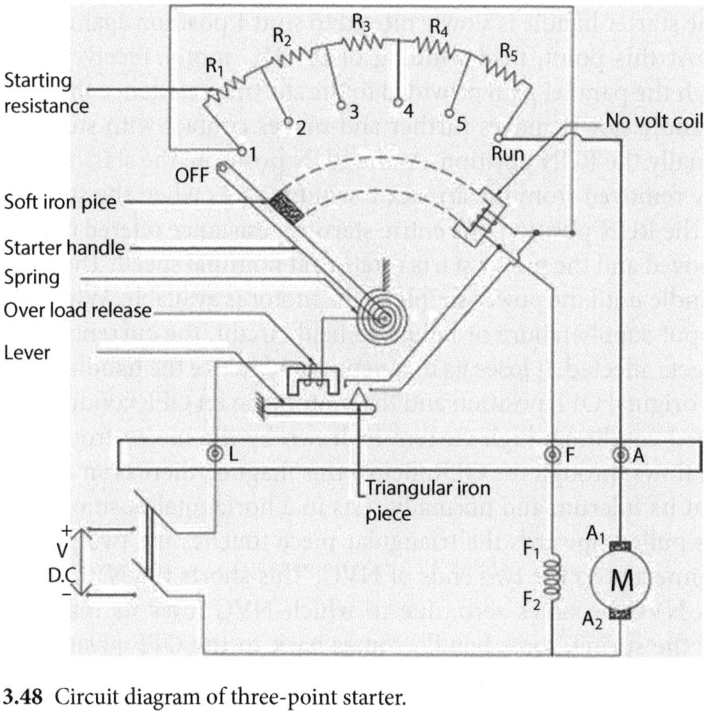


**Figure 3.48** Circuit diagram of three-point starter.

The starter basically has variable resistances and this variable resistance is divided into multiple numbers of positions. The contact points of these resistance positions, called studs, are taken out separately and named as OFF, 1, 2, 3 .... RUN as shown in Figure 3.48. A three-point starter has three main points, which are "L", "F" and "A".

Where,

L" is line terminal of supply positive
"A" is connected to the armature winding of the DC motor
"F" is connected to the field winding of the DC motor

Point "L" is connected to the Overload Release (OLR) and the other end of the OLR is connected to a point where handle of the starter is pivoted. This handle is movable from the OFF position to RUN positions freely against the spring force. The handle can come to the original OFF position due to the force of the spring. The starting resistance is fully in series with the armature winding of the DC motor. The power supply for No Volt Coil (NVC) is taken from stud "1" and it is connected to terminal 'F'.

Operation of three-point starter:

Initially the starter handle position is OFF. The DC supply is switched ON and the starter handle is slowly moved to stud 1 position against the spring force. At this point, field winding of the DC motor receives the supply through the parallel path provided to the starting resistance through NVC. The handle slowly moves further and makes contact with studs 2, 3, 4, 5 and finally the RUN position. At the RUN position, the starting resistance is fully removed from the armature winding. i.e., when the starter handle keeps the RUN position, the entire starting resistance offered by the starter is removed and the motor starts rotating at nominal speed. The NVC holds the handle until the power supply to the motor is available. Whenever there is a input supply failure or failure of field circuit, the current through the NVC gets affected. It loses its magnetism and hence the handle comes back to the original OFF position and the motor is in an OFF condition. During overload condition, high current is drawn by the motor from the supply which flows through the OLR. Below this magnet, there is an arm which is fixed at its fulcrum and normally rests in a horizontal position. When the arm is pulled upwards the triangular piece touches the two points which are connected to the two ends of NVC. This shorts the NVC and voltage across NVC becomes zero, due to which NVC loses its magnetism. So under the spring, force handle comes back to the OFF position, disconnecting the motor from the supply.

### B. Four-point starter

The basic difference between three-point and four-point starter is connection of NVC. In three-point, NVC is in series with the field winding while in a four-point starter NVC is connected separately across the power supply through the fourth terminal called 'N' in addition to the terminals 'L', 'F' and 'A' as shown in Figure 3.49.

In a four-point starter, any change in the field current does not affect the NVC and also ensures the NVC always produces a force to hold the handle against the spring force at RUN position.

### C. Two-point starter

In a DC series motor, field winding and armature windings are connected in series and hence starting resistance is inserted in series with the field and armature winding. The starter used to limit the starting current for DC series motor is called a two-point starter. The circuit diagram of two-point starter for DC series motor is shown in Figure 3.50.

It has only two terminals, namely Line (L) and Field (F). The "F" terminal is one end NVC and the other end is connected at the series combination of field and the armature winding. When the handle is moved to stud 1, the motor gets supply and the entire starting resistance is in series with the armature and field winding, hence starting current is limited. The NVC

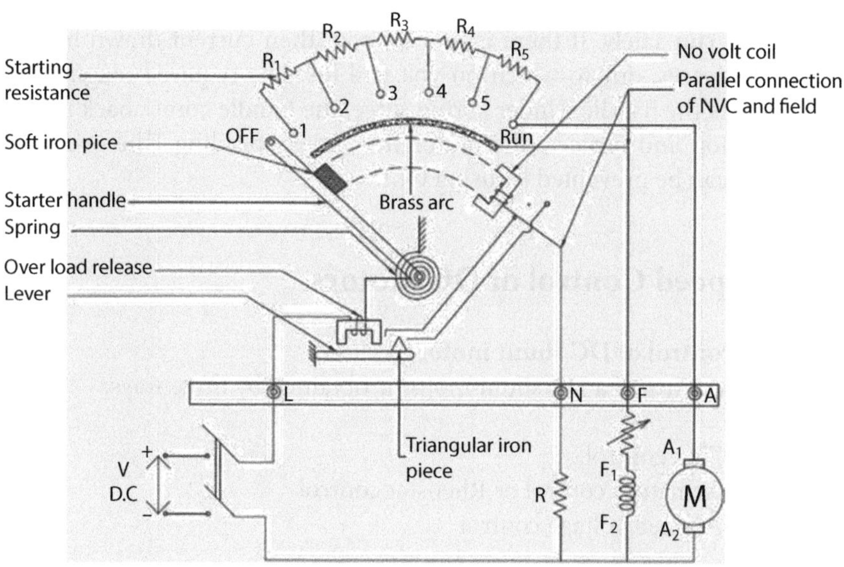


**Figure 3.49** Circuit diagram of four point starter.

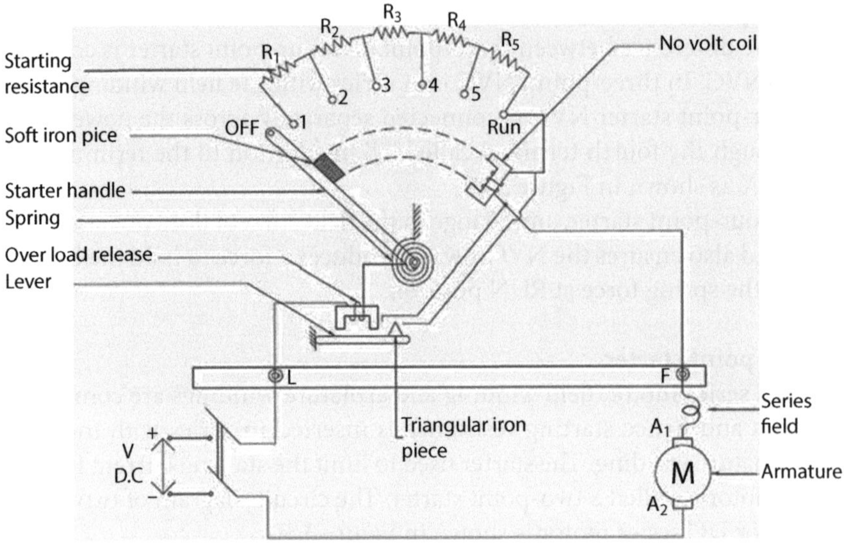


**Figure 3.50** Circuit diagram of two-point starter.

holds the handle by attracting the soft iron piece on the handle. The no volt coil is also called hold on coil. The main disadvantage of a DC series motor is over-speeding action when the load is less or no load; this can be prevented by using the tw -starter. The NVC is designed to hold the handle in the RUN position only when it carries sufficient current, for which the motor can run safely. If there is loss of load, then current drawn by the motor decreases, due to which no volt coil loses its required magnetism and releases the handle. Under spring force, the handle comes back to the OFF position and protects the motor from over-speeding. The overload condition can be prevented by using OLR.

## 3.11 Speed Control of DC Motors

**A. Speed control of DC shunt motor**

The speed control of a DC shunt motor is obtained by three ways:

i. Flux control
ii. Armature control or Rheostat control
iii. Applied voltage control

i. Flux Control

The speed of the DC motor is controlled by varying the flux produced by the field winding. This method is called flux control. The circuit diagram for flux control-based DC shunt motor speed control is shown in Figure 3.51.

The speed is inversely proportional to the current in the flow field winding as shown in Figure 3.52.

The flux is dependent on the current through the shunt field winding. Thus flux can be controlled by adding a rheostat in series with the shunt field winding as shown in Figure 3.51. Initially, rheostat (R) is kept at

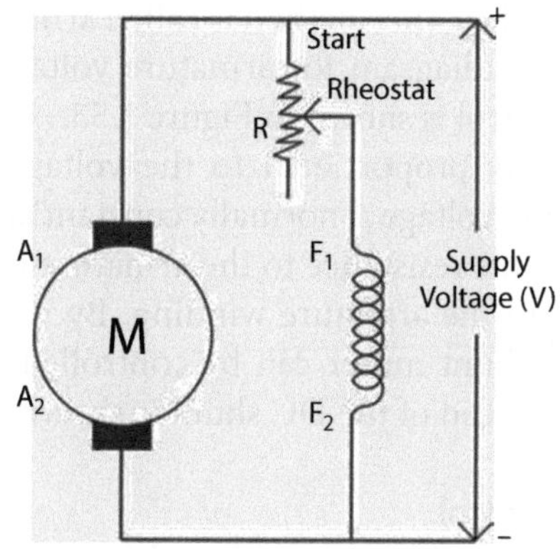


**Figure 3.51** Circuit diagram for flux control-based DC shunt motor speed control.

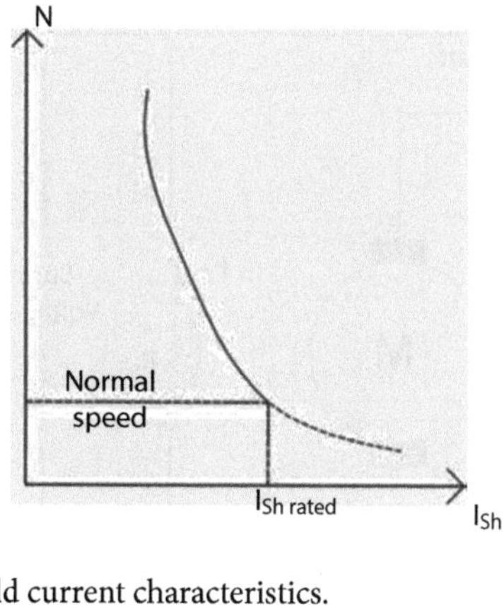


**Figure 3.52** Speed vs. field current characteristics.

minimum position and the full input voltage is supplied to the field winding, so current through shunt field winding is at its rated value. Hence the motor is rotating at its rated speed. Then, rheostat (R) position is increased due to which shunt field current $I_{sh}$ is decreases and a resulting decrease in field flux is produced. The relationship between the speed and flux is expressed in equation 3.35.

$$N\alpha\frac{1}{\phi} \tag{3.35}$$

ii. Armature voltage or Rheostat control method

The speed of the DC motor is controlled by varying the flux produced by the armature winding [12]. This method is called armature voltage or rheostat control. The circuit diagram for armature voltage control-based DC shunt motor speed control is shown in Figure 3.53.

The speed is directly proportional to the voltage applied across the armature. As the supply voltage is normally constant and speed is also constant, adding the variable resistance to the armature winding can vary the input supply voltage to the armature winding. By varying the resistance, the speed of the DC shunt motor can be controlled. The relationship of armature voltage and speed of the DC shunt is shown in Figure 3.54.

iii. Applied Voltage Control

The circuit diagram for applied voltage-based DC shunt motor speed control is shown in Figure 3.55.

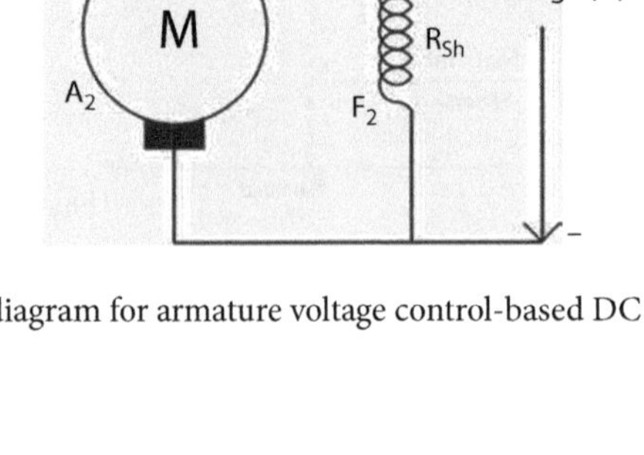


**Figure 3.53** Circuit diagram for armature voltage control-based DC shunt motor speed control.

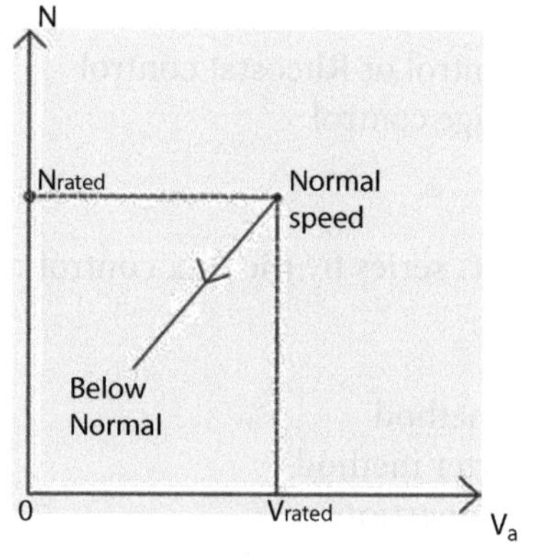


**Figure 3.54** Relationship of armature voltage and speed.

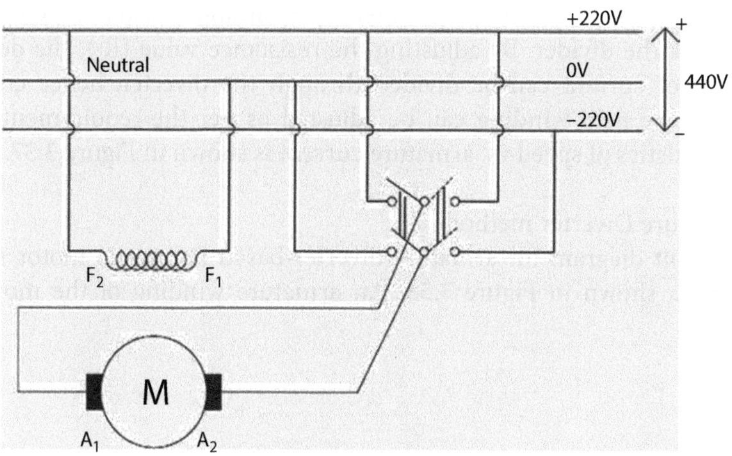


**Figure 3.55** Circuit diagram for applied voltage-based DC shunt motor speed control.

The shunt field winding of the motor is permanently connected to the fixed DC input voltage supply as shown in Figure 3.55. The armature winding is supplied with two voltage level by means of suitable switching arrangements. In one position, 220 V is supplied to armature winding and in another position 440 V is supplied to the armature winding.

**B. Speed control of DC series motor**

The speed control of DC shunt motor is obtained by three ways:

i. Flux control
ii. Armature control or Rheostat control
iii. Applied voltage control

i. Flux control

The speed control of DC series by the flux control method is obtained by four types:

a. Field diverter method
b. Armature diverter method
c. Tapped field connection
d. Series-parallel connection of field

a. Field diverter method

The circuit diagram for field diverter-based DC series motor speed control is shown in Figure 3.56.

The series field winding is shunted by a variable resistance ($R_x$) and is known as the divider. By adjusting the resistance value ($R_x$), the desired amount of current can be divided through the diverter, hence current through the field winding can be adjusted as per the requirement. The characteristics of speed vs. armature current is shown in Figure 3.57.

b. Armature Diverter method

The circuit diagram for armature diverter-based DC series motor speed control is shown in Figure 3.58. An armature winding of the motor is

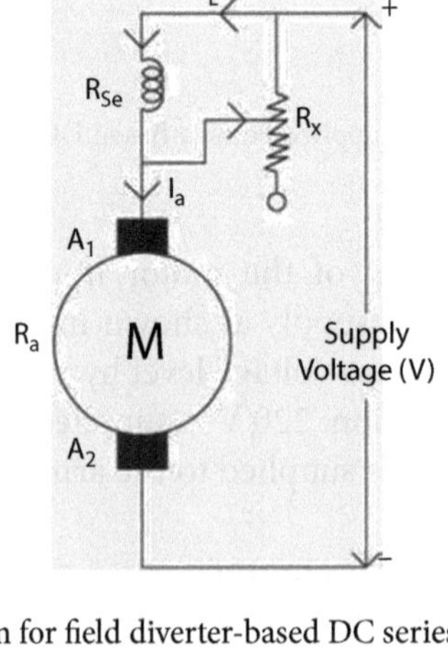


**Figure 3.56** Circuit diagram for field diverter-based DC series motor speed control.

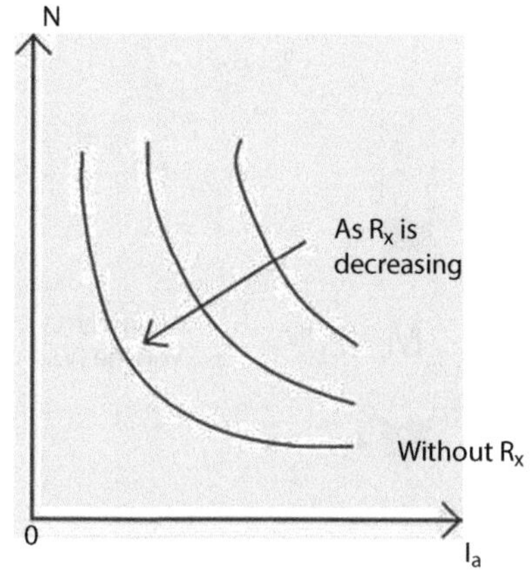


**Figure 3.57** Characteristics of speed vs. armature current.

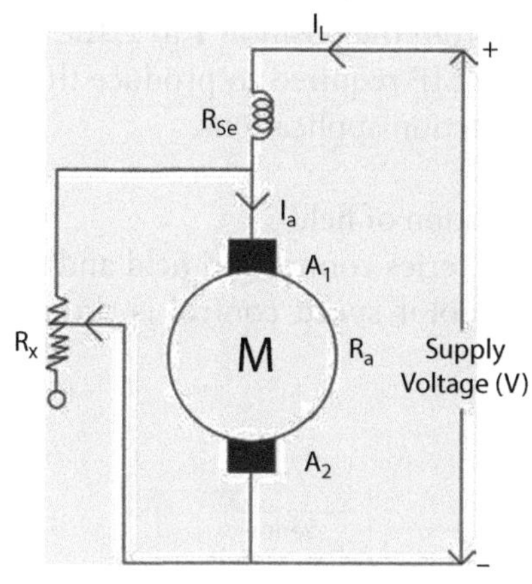


**Figure 3.58** Circuit diagram for armature diverter-based DC series motor speed control.

shunted with an external variable resistance ($R_x$). By adjusting the resistance value ($R_x$), the desired amount of current can be divided through the diverter, hence current through the armature winding can be adjusted as per the requirement.

c. Tapped field method

The circuit diagram for tapped field-based DC series motor speed control is shown in Figure 3.59.

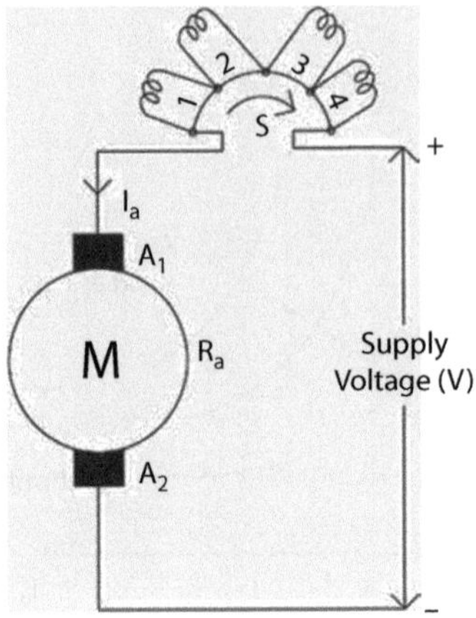


**Figure 3.59** Circuit diagram for tapped field-based DC series motor speed control.

Flux change is achieved by changing the number of turns of the field winding. Switch "S" is used to select the number of turns. The number of turns can be increased from the position 1 to 2.The number of field winding selected based on MMF required to produce the flux. This method is often used in electric traction applications.

d. Series-Parallel connection of field

The circuit diagram for series connection field and parallel connection of field-based DC series motor speed control is shown in Figure 3.60 and Figure 3.61, respectively.

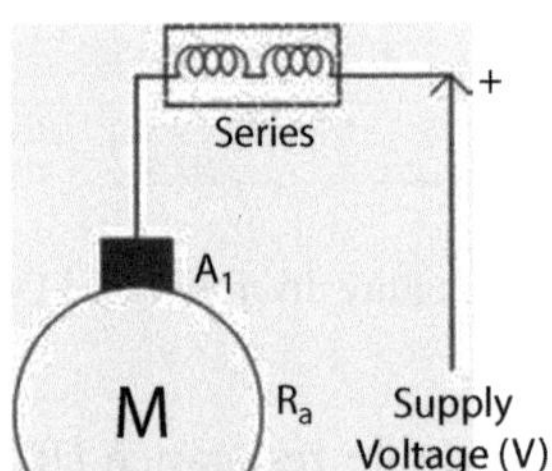


**Figure 3.60** Circuit diagram for series connected field-based DC series motor speed control.

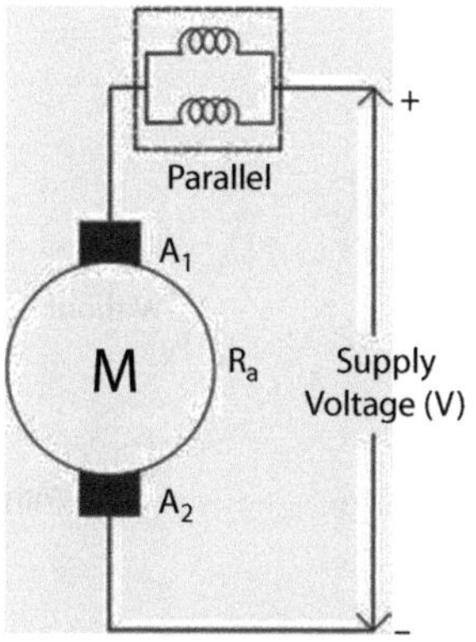


**Figure 3.61** Circuit diagram for parallel connected field-based DC series motor speed control.

For producing the same torque, the field winding can be arranged in series or parallel. Based on the connection, MMF produced by the winding changes results in flux changes. Hence, speed can be controlled.

ii. Rheostat Control

The circuit diagram for rheostat control-based DC series motor speed control is shown in Figure 3.62.

The variable resistance ($R_x$) is connected series with the motor circuit as shown in Figure 3.62 and voltage drop across the variable resistance is $I_aR_x$. By varying the resistance, voltage across the armature winding is reduced.

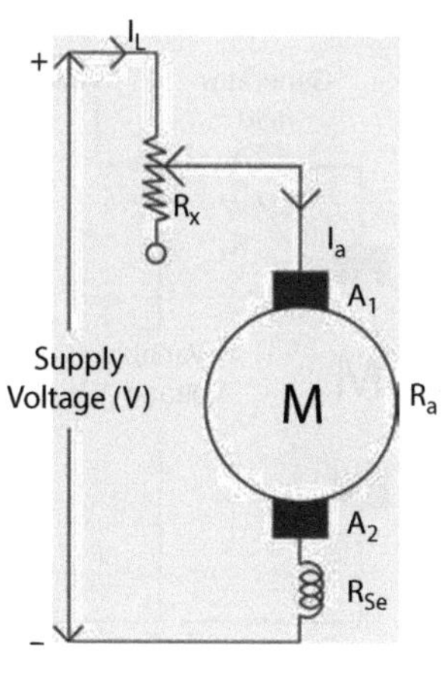


**Figure 3.62** Circuit diagram for rheostat-based DC series motor speed control.

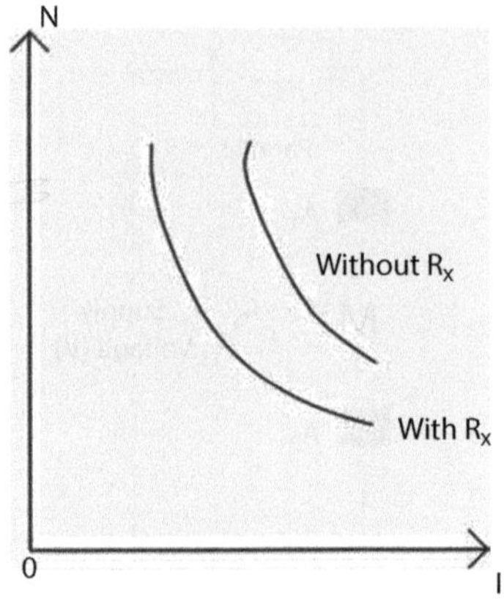


**Figure 3.63** Relationship between speed vs. armature current.

Hence, speed of the DC motor is controlled. The relationship between speed vs. armature current is shown in shown in Figure 3.63.

iii. Applied voltage control

The circuit diagram of applied voltage-based speed control of DC series motor is shown in Figure 3.64.

The DC series motor is controlled by external generator. The generator is driven by a suitable prime mover. The output voltage from the generator is controlled by a field diverter resistance connected across the generator series field winding. By varying the generator output voltage, the speed of the DC series motor is controlled.

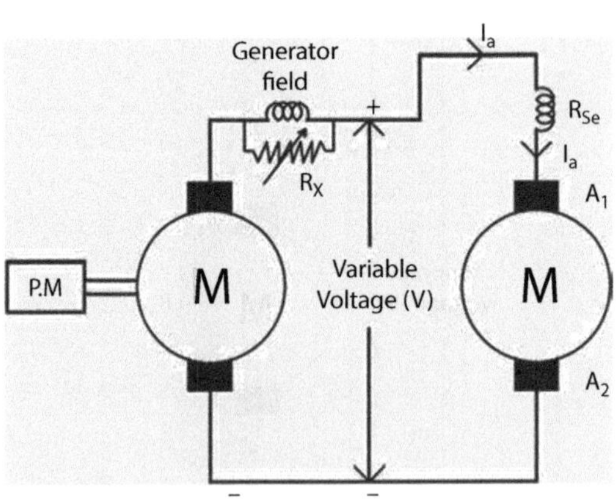


**Figure 3.64** Circuit diagram of applied voltage-based speed control of DC series motor.

## 3.12 Universal Motor

A universal motor is a motor which can be operated either by DC supply or single Phase AC Supply. By construction, a universal motor is the same as the DC series motor. These motors are generally series wound (armature winding and field winding are in series) and it produces the higher starting torque. Generally, a universal motor comes inbuilt with the devices they are meant for. Most of the universal motors are designed to operate at higher speeds and exceeding 3500 RPM. They run at a lower speed on AC supply as compared with DC supply of same voltage, due to voltage drop by reactance.

**A. Construction of Universal motor**

The construction of a universal motor is shown in Figure 3.65 and it is similar to DC machine.

It consists of a stator on which field poles are mounted. Field windings are wound on the field poles. Like DC machines, both stator and rotor

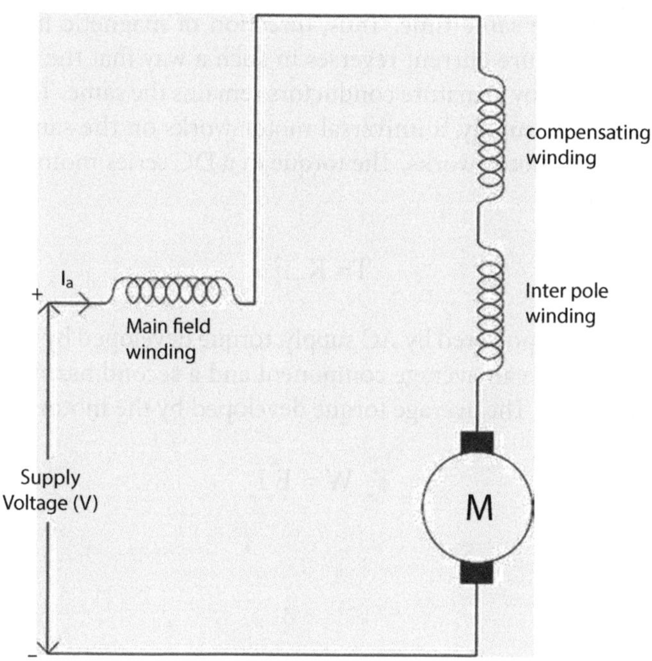


**Figure 3.65** Construction of universal motor.

are laminated. Laminations are used to minimize the eddy currents which induce while the universal motor operating on AC supply. The rotating armature winding is wound type and has straight or skewed slots and commutator with brushes resting on it. The commutation on AC supply is poor as compared with DC supply, because of the current induced in the armature coils. For that reason brushes are used that have high resistance.

**B. Working of Universal Motor**

A universal motor works either by DC or single-phase AC power supply. When the universal motor is powered by DC supply, it works as a DC series motor. When current flows in the field winding, it produces an electromagnetic field. The same current flows through the armature winding. When a current carrying conductor is placed in an electromagnetic field, it experiences a mechanical force and direction of force according to Fleming's left hand rule. Due to this mechanical force, the rotor starts rotating.

When the universal motor is powered by AC supply, it produces the unidirectional torque. Because both armature winding and field winding are connected in series, they are in same phase. Hence, as the polarity of AC changes periodically, the direction of current in armature and field winding reverses at the same time. Thus, direction of magnetic field and the direction of armature current reverses in such a way that the direction of force experienced by armature conductors remains the same. Thus, regardless of AC or DC supply, a universal motor works on the same principle that a DC series motor works. The torque in a DC series motor is given in equation 3.36.

$$T = K_T i_a^2 \tag{3.36}$$

If this motor is powered by AC supply, torque developed by the motor is unidirectional with an average component and a second harmonics oscillating component. The average torque developed by the motor is

$$T_{av} W = E_a I_a$$

**Students and Lecturer conversation**

1. Lecturer: What is a Generator?
**Student:** A Generator is an electromechanical device which is used to convert Mechanical Energy into Electrical Energy.
2. Lecturer: On which principle does a Generator work?
**Student:** Generator works on the principle of Faraday's laws of "Electro Magnetic Induction".

3. Lecturer: What are the significant parts of a DC Generator?
**Student:** The significant parts of a DC generator are

1. Stator
2. Rotor
3. Field winding
4. Armature Winding
5. Commutator and brushes

4. Lecturer: What are the characteristics of a DC Generator?
**Student:** The characteristics of a DC generator are

1. Magnetizing Characteristics
2. Load Characteristics

5. Lecturer: What are the two main parts of a DC Generator?
**Student:** The two main parts are

1. Stator
2. Rotor

6. Lecturer: Where is wave winding used?
**Student:** Wave winding is used in high voltage and low current DC machines.

7. Lecturer: Where is lap winding used?
**Student:** Lap winding is used in low voltage and high current DC machines.

8. Lecturer: What is the effect of armature reaction in a DC machine?
**Student:** The effect of armature reaction is to decrease the total flux and hence it reduces the voltage or EMF.

9. Lecturer: How many commutator segments will be in a 24-coil DC machine?
**Student:** 24 commutator segments. Because in a DC machine, the number of commutator segments are equal to the number of coils.
10. Lecturer: How is the size of a DC generator reduced?
**Student:** The size of a DC generator can be reduced by using magnetic material of high permeability.

11. Lecturer: What is the main difference between an AC Generator and a DC Generator?
**Student:** In an AC Generator the field is rotating and the armature is stationary, whereas in a DC Generator the field is stationary and the armature is rotating.

12. Lecturer: What is the role of the commutator?
**Student**: The commutator is like a rectifier, it converts the AC supply into the DC supply within the armature winding to across the brushes. It is essential because EMF induced in the armature conductor is alternating current.

13. Lecturer: What is the function of brushes?
**Student:** The electrical connection between the commutator and external cicuit is achieved with the help of brushes.

14. Lecturer: What are the different types of DC generators?
**Student:** Generators are generally classified based on their methods of field excitation, which are as follows:

(i) Separately excited DC generators
(ii) Self-excited DC generators

## References

1. D. P. Kothari and I J Nagrath, *Electric Machines*, 4th Edison, Tata McGraw Hill, New Delhi, 2010.
2. S. K. Sahdev, *Electric Machines*, Cambridge University Press, New Delhi, 2018.
3. Saurabh Kumar Mukerji, Ahmad Shahid Khan and Yatendra Pal Singh, *Electromagnetics for Electrical Machines*, CRC Press, Boca Raton, 2015.
4. B. L. Theraja and A. K. Theraja, *Electrical Technology*, volume 2, S. Chand publication, 2006.
5. Jacek F. Gieras, *Electrical Machines*, CRC press, Boca Raton, 2017.
6. Prithwiraj Purkait and Indrayudh Bandyopadhyay, *Electrical Machines*, Oxford Publications, First edition, New Delhi, 2017.
7. Melkebeek Jan, *Electrical Machines and Drives*, Springer, Singapore, 2018.
8. R. K. Rajput, *Electrical Machines*, Laxmi Publications, Fourth Edition, New Delhi, 2006.
9. Theodore Wildi, *Electrical Machines, Drives, and Power Systems*, Sixth Edition, Pearson Publication, Essex, 2014.
10. Seung-Ki Sul, *Control of Electric Machine Frive Systems*, IEEE Press, United State of America, 2011.
11. Juha Pyhonen, Tapani Jokinen, Valeria Hrabovcoa and Hanna Niemela, *Design of Rotating Electrical Machines*, John Wiley & Sons, United Kingdom, 2008.
12. Slobodan N. Vukosavic, *Electrical Machines*, Springer, New York, 2013.

# 4
# AC Machines

## 4.1 Introduction

An electric motor is a device which is used to convert the electrical energy into rotating mechanical energy. This rotating mechanical energy is used for various applications like fans, drives, water pumping, etc. The motors that operate in DC power supply are called DC motors and the motors operating in AC power supply are called AC motors [1, 17].

Based on the operation, AC motors are classified into the following types:

i) Induction motors
ii) Synchronous motors
iii) Special purpose motors

Based on the number of phases, AC motors are classified into the following types:

i) Single-phase motors
ii) Three-phase motors

## 4.2 Three-Phase Induction Motor

Three-phase induction motors are widely used for various residential, commercial, industrial and agricultural applications [4, 6]. The advantages of three-phase induction motors over other types of motors are:

i) Induction motors have a self-starting property and no need for external starting devices
ii) High power factor
iii) Simple and rugged construction

P. Sivaraman, C. Sharmeela, A. Thaiyal Nayagi, and R. Mahendran. Basic Electrical and Instrumentation Engineering, (185–242) © 2021 Scrivener Publishing LLC

iv) High reliability
v) Efficiency is high
vi) Low cost
vii) Good speed regulation
viii) Less maintenance is required

The three-phase induction motor works on the principles of Rotating Magnetic Field (RMF).

### 4.2.1 Rotating Magnetic Field

The RMF is defined as the magnetic field or flux having constant amplitude and its axis continuously rotating in a plane with a certain speed. When a balanced three-phase voltage is supplied to a balanced three-phase winding of an electric motor, it produces RMF at constant magnitude [2]. This RMF field is also called synchronous speed.

### 4.2.2 Construction

Basically, the induction motor consists of two main parts, namely:

i) Stator – stationary part of the motor
ii) Rotor – rotating part of the motor

The conversion of electrical energy into rotating mechanical energy takes place in the rotor. Hence the rotor develops a driving torque and rotates.

**A. Stator**

The stator has laminated types of construction and is made up of a multiple number of stamping whose thickness ranges from 0.4 to 0.5 mm. The stampings are slotted in its periphery to carry the stator winding and are insulated from each other [3]. So this type of construction reduces the iron losses to a minimum value. The typical construction of stator is shown in Figure 4.1.

The typical stator of an induction motor is shown in Figure 4.2.

The stampings are made together to the core of the stator and then fitted in a frame. The choice of material for the stamping is generally silicon steel, which minimizes the hysteresis loss. The three-phase winding is wound on the stator slots and connected either in star or delta connection. When this winding is excited by a three-phase supply it will produce the RMF. The

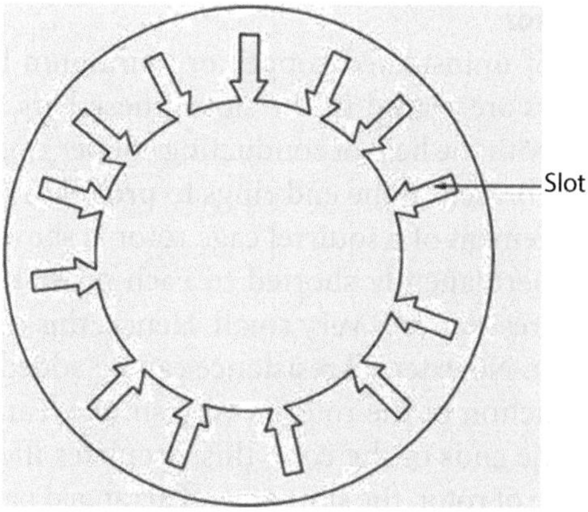


**Figure 4.1** Construction of stator.

**Figure 4.2** Typical view of stator of an induction motor.

selection of number of poles depends on the speed of the RMF. The radial ducts are provided for the cooling purpose.

## B. Rotor

The rotor core is laminated in construction and made up of cast iron. It is cylindrical, with slots on its periphery and placed inside the stator [5].

Based on the construction of the rotor, it is classified into two types:

i) Squirrel cage rotor
ii) Wound or slip ring rotor

**i. Squirrel Cage Rotor**

The rotor consists of uninsulated copper or aluminum bars called rotor conductors. The bars are placed in the slots. These bars are permanently shorted at each end with the help of conducting copper ring called end ring. The bars are usually brazed to the end rings to provide a good mechanical strength. The arrangement of a squirrel cage rotor is shown in Figure 4.3.

As the bars are permanently shorted to each other through end ring and the entire rotor resistance is very small. Hence, this rotor is also called short circuited Rotor. No external resistance can be added in the rotor circuit. So, the construction of the rotor is very simple. Fan blades are generally provided at the ends of the core; this circulates the air through the machine. In this type of rotor, the slots are not arranged parallel to the shaft axis but are skewed as shown in Figure 4.4.

The typical skewed arrangement of a squirrel cage induction motor is shown in Figure 4.5.

The advantages of skewing the rotor bars are listed below:

i) It avoids the magnetic locking between the stator and rotor
ii) A magnetic hum, i.e., noise gets reduced to skewing

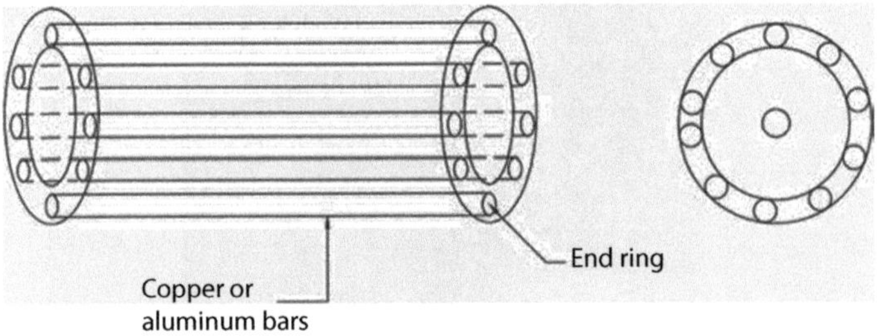


**Figure 4.3** Arrangement of squirrel cage rotor.

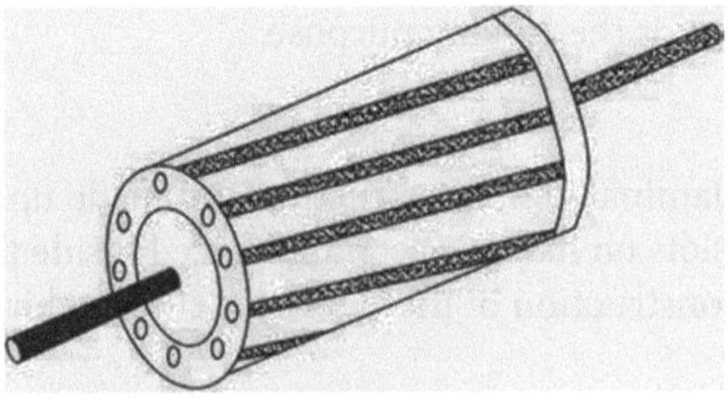

**Figure 4.4** Skewed arrangement of squirrel cage rotor.

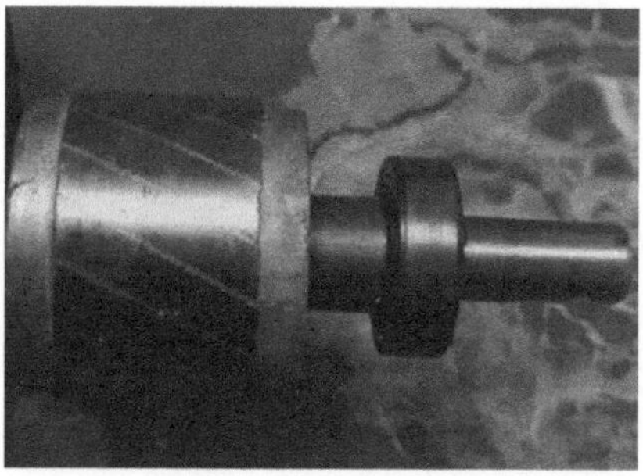

**Figure 4.5** Typical skewed arrangement of squirrel cage induction motor.

iii) It makes motor operation smooth
iv) It increases the effective transformation ratio between stator and rotor

**ii. Slip Ring or Wound Rotor**

The rotor winding is exactly similar to the stator winding. The slot in the rotor contains the rotor winding. The three ends of the three-phase windings available after connecting the winding in either star or delta are permanently connected to the slip rings. So in this type of rotor, the external resistances can be added with the help of brushes and slip-ring arrangement, in series with each phase of the rotor winding [1, 4].

Figure 4.6 shows the slip ring rotor with external star connected resistance.

At the same time brushes are also lifted from the slip rings. This avoids wear and tear of the brushes due to friction.

The typical slip ring induction motor is shown in Figure 4.7.

The typical slip ring and carbon brush arrangement is shown in Figure 4.8.

## 4.2.3 Working Principle

An induction motor works on the principle of Faraday's law of electromagnetic induction. When the three-phase AC power supply is given to the stator winding of a three-phase induction motor, it produces the RMF at constant magnitude. The speed of the rotating magnetic field is given in equation 4.1.

$$N_S = \frac{120F}{P} \tag{4.1}$$

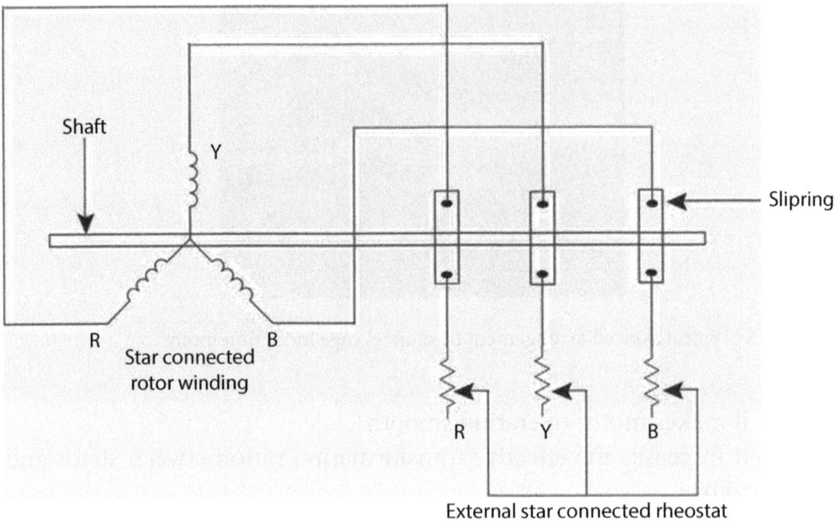


**Figure 4.6** Slip ring rotor.

**Figure 4.7** Typical slip ring induction motor.

Where,

F is frequency of the supply voltage

P is number of pole

According to Faraday's law of electromagnetic induction, whenever conductor cuts the rotating magnetic flux, EMF gets induced in it. The induced EMF in rotor conductor is called rotor induced EMF and the current that circulates through the rotor is called rotor current [3]. The current flow in the rotor circuits that produces the flux is called rotor flux. Now there are two fluxes available, namely stator flux (RMF) and rotor flux. The rotor

**Figure 4.8** Typical slip ring and carbon brush arrangement.

conductor is shown in Figure 4.9 (a), current induced in rotor circuit is shown in Figure 4.9 (b) and flux due to induced current in the rotor is shown in Figure 4.9 (c).

As the flux lines act as stretched rubber band, high flux density area exerts a push on rotor conductor towards low flux density area. Hence, the rotor conductor experiences a force from high magnetic flux. As the rotor conductors experience a force, the overall rotor experiences a torque and starts rotating. So interaction of the two fluxes is essential for a motoring action. As per Lenz's law, the direction of current induced in the rotor circuit opposes the cause producing it.

The direction of current induced in the rotor circuit is shown in Figure 4.10.

$N_S$ is speed of RMF in RPM

N is speed of rotor in RPM

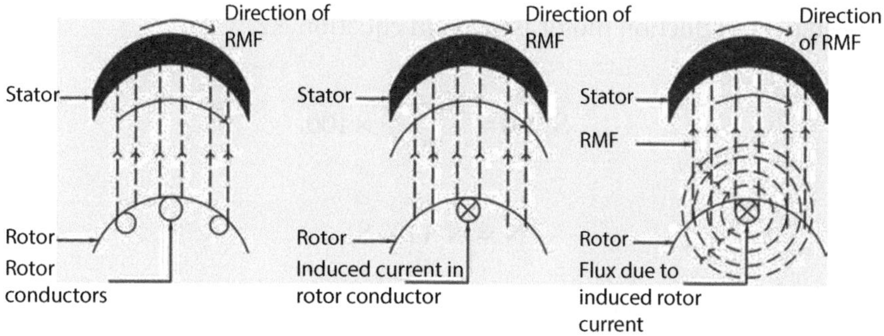


**Figure 4.9** Direction of current flow in rotor circuit.

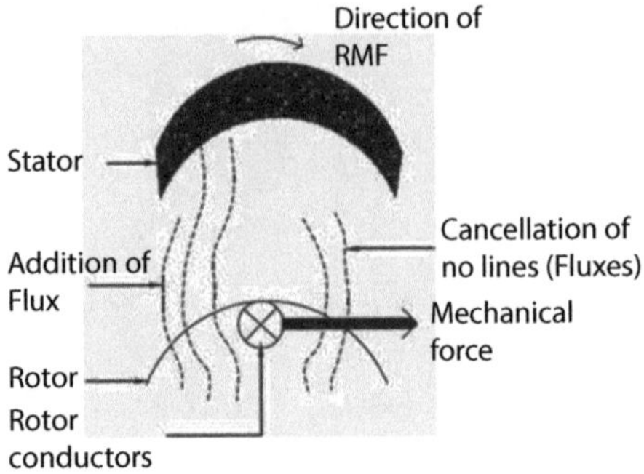


**Figure 4.10** Direction of current flow in rotor circuit.

$N_S$ – N is relative speed between two rotating magnetic field and the rotor conductor

Hence, the rotor always rotates in the same direction as that of the RMF. When the rotor starts rotating, it tries to rotate at the same speed of the RMF. If the rotor rotates at the speed of the RMF, the relative motion between rotor and RMF is zero ($N_S - N = 0$). There is no induced EMF in the rotor, so current flow in the rotor is zero and hence torque produced by the motor is zero. Therefore, the rotor rotates at slightly less speed than synchronous speed. It means, the rotor of an induction motor always rotates at less speed than synchronous speed. Hence this motor is called as asynchronous motor.

### 4.2.4 Slip of an Induction Motor

The speed difference between the synchronous speed ($N_S$) and actual speed of rotor (N) is called slip speed or slip of an induction motor [2]. The slip speed of an induction motor is given in equation 4.2.

$$S\ (\%) = \frac{N_S - N}{N} \times 100 \tag{4.2}$$

$$N = N_s\ (1 - S)$$

During the starting time, S = 1, so N = 0

Under no load condition, the difference of speed between the actual rotor speed (N) and synchronous speed ($N_S$) is about 1%.

The effects of slip on rotor parameters are as follows:

- Rotor Frequency
- Magnitude of rotor induced EMF
- Rotor reactance
- Rotor power factor
- Rotor current

### 4.2.5 Torque Equation

The torque developed by the induction motor is based on the factors listed below:

1. The induced EMF in the rotor circuit
2. The rotor current magnitude value
3. The power factor of the rotor circuit

The mathematical relationship for torque of an induction motor can be expressed in equation 4.3.

$$T \propto \phi I_{2r} \, Cos\phi_{2r} \tag{4.3}$$

Where,

$\Phi$ is flux to produce induced EMF in rotor
$I_{2r}$ is rotor running Current
$Cos\Phi_{2r}$ is rotor power factor

The flux ($\Phi$) produced by stator is directly proportional to supply voltage ($E_1$) and given equation 4.4.

$$\phi \propto E_1 \tag{4.4}$$

While $E_1$ and $E_2$ are related to each other through ratio of stator turns to rotor turns and given in equation 4.5.

$$\frac{E_1}{E_2} = K \tag{4.5}$$

Using equation (4.4) and (4.5),

$$E_2 \propto \phi \tag{4.6}$$

Substitute equation (4.6) in (4.3)

$$T \propto E_2 I_{2r} \operatorname{Cos}\phi_{2r}$$

While,

$$I_{2r} = \frac{E_{2r}}{Z_{2r}} = \frac{SE_2}{\sqrt{R_2^2 + \left(SX_2^2\right)}} \tag{4.7}$$

$$\operatorname{Cos}\phi_{2r} = \frac{R_2}{Z_{2r}} = \frac{R_2}{\sqrt{R_2^2 + \left(SX_2^2\right)}} \tag{4.8}$$

The relation of the rotor, resistance and reactance is shown in Figure 4.11.

Using equation (4.6), (4.7) and (4.8) the equation (4.3) can be written as

$$T \propto E_{2r} \times \frac{SE_2}{\sqrt{R_2^2 + \left(SX_2^2\right)}} \times \frac{R_2}{\sqrt{R_2^2 + \left(SX_2^2\right)}}$$

$$T \propto \frac{SE_2^2 R_2}{\sqrt{R_2^2 + \left(SX_2\right)^2}} \text{ N} - \text{m}$$

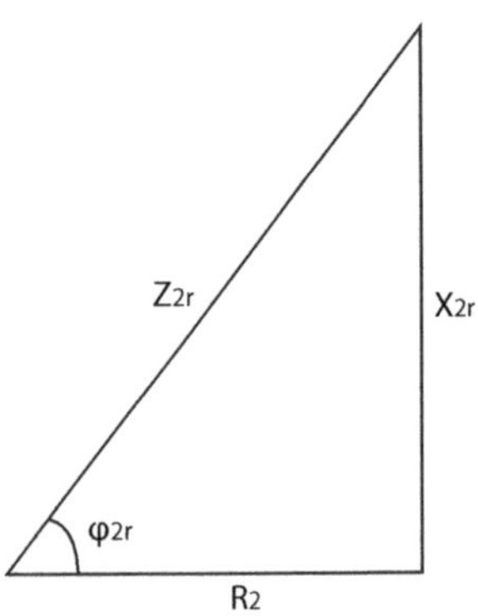


**Figure 4.11** The relation of the rotor, resistance and reactance.

$$T = \frac{KSE_2^2R_2}{\sqrt{R_2^2 + (SX_2)^2}} \text{ N-m} \tag{4.9}$$

Where,

K is constant of proportionality

The constant 'K' is proved to be $\frac{3}{2\pi N_s}$ for the three-phase induction motor.

$$K = \frac{3}{2\pi N_s} \tag{4.10}$$

$N_S$ is synchronous speed in RPS = $\frac{N_s}{60}$ v

Sub equation (4.10) in (4.9)

$$T = \frac{3}{2\pi N_s} \times \frac{SE_2^2R_2}{\sqrt{R_2^2 + (SX_2)^2}} \text{ N-m} \tag{4.11}$$

The torque developed at any load condition can be obtained if slip at the load is known and standstill rotor parameters are known.

### 4.2.6 Torque–Slip Characteristics

The speed of the induction motor decreases with increase of loading from no load to full load resulting in an increase of slip speed or slip. The curve obtained by plotting torque against slip is called torque – slip characteristics of an induction motor, i.e., plotting the slip value from zero at synchronous speed to 1. The Torque–Slip characteristics of an induction motor are shown in Figure 4.12.

The torque equation of the three-phase induction motor is given in equation (4.12).

$$T \propto \frac{SE_2^2R_2}{R_2^2 + (SX_2)^2} \tag{4.12}$$

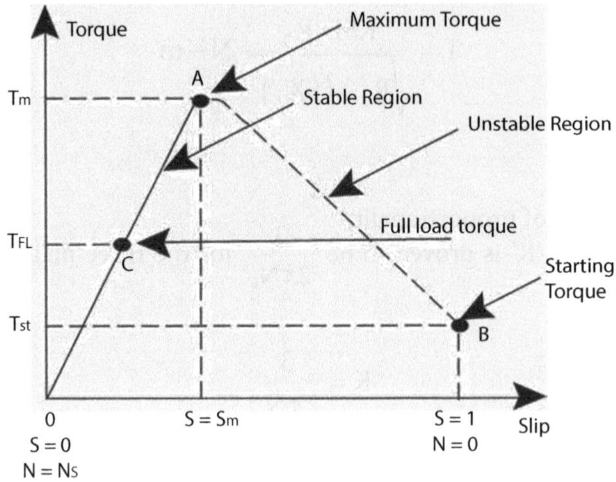


**Figure 4.12** Torque–Slip characteristics of an induction motor.

The input supply voltage is constant, so torque equation (4.12) can be rewritten as

$$T \propto \frac{SR_2}{R_2^2 + (SX_2)^2} \tag{4.13}$$

**Low slip region:** When the slip 'S' is very less, $SX_2^2$ also very less as compared with $R_2^2$ and it can be ignored. The torque equation (4.13) can be rewritten as

$$T \propto \frac{SR_2}{R_2^2} \propto S \tag{4.14}$$

Where,

$R_2$ is constant

This low slip region is called stable region and torque in this region is directly proportional to slip. If slip increases then torque also increases and if slip decreases then torque also decreases.

**High slip region:** When the slip 'S' is high, between $S_m$ and 1, this region is called unstable region. The $R_2^2$ value is small as compared with $SX_2^2$. The torque equation (4.15) can be rewritten as

$$T \propto \frac{SR_2}{(SX_2)^2} \propto \frac{1}{S} \tag{4.15}$$

Where,

$R_2$ and $X_2$ are constant

This torque is inversely proportional to slip. If slip increases then torque decreases and if slip decreases then torque increases.

### 4.2.7 Induction Motor as a Transformer

The transformer is a device having two windings which are magnetically connected and physically disconnected. When one winding is powered by AC supply of particular frequency, the EMF is induced in the second winding at the same frequency of power supply. The winding, in which power supply is given, is called primary winding and the winding in which EMF gets induced is called secondary winding. The induction motor also having two windings and principle of induction motor is the same as a transformer, i.e., energy transferred from the stator to the rotor through the magnetic flux. Hence, the induction motor is called a rotating transformer. The primary and secondary winding of the three-phase induction motor is shown in Figure 4.13.

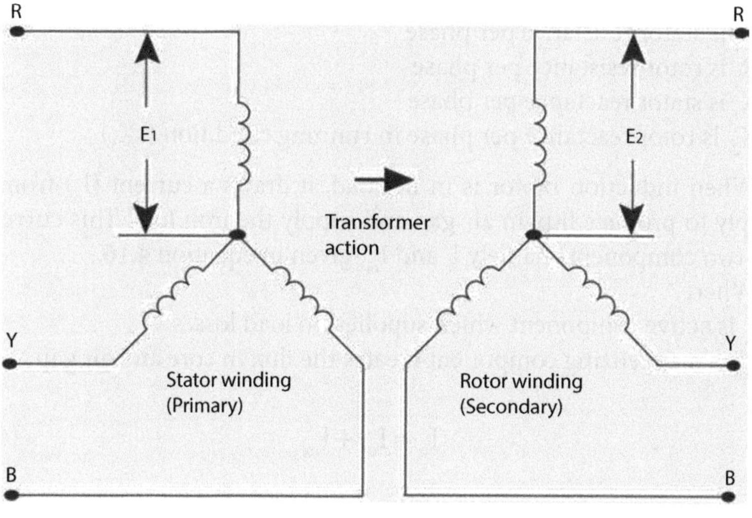


**Figure 4.13** Primary and secondary winding of the three-phase induction motor.

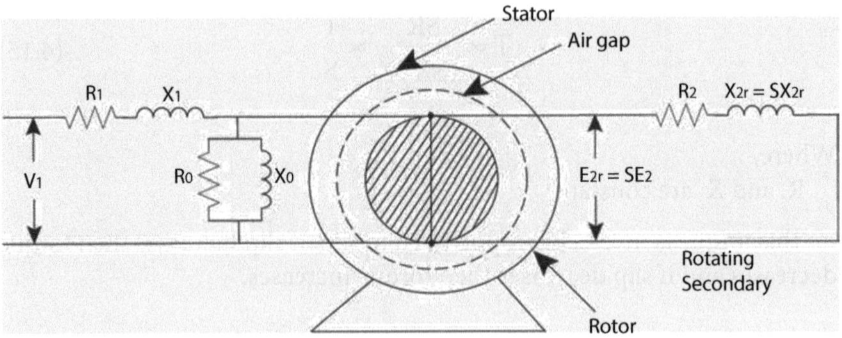


**Figure 4.14** Circuit diagram of an induction motor as transformer.

The circuit diagram of an induction motor can be represented as a transformer as shown in Figure 4.14.

## 4.2.8 Equivalent Circuit of Induction Motor

$E_1$ is stator induced EMF per phase (due to self induction)

$E_2$ is rotor induced EMF at standstill condm nition (due to mutual induction)

$$k = \frac{\text{Rotor turns}}{\text{Stator turns}} = \frac{E_2}{E_1} \text{ is transformation ratio}$$

$E_{2r}$ is rotor induced EMF at running condition

$R_1$ is stator resistance per phase

$R_2$ is rotor resistance per phase

$X_1$ is stator reactance per phase

$X_{2r}$ is rotor reactance per phase in running condition ($sX_2$)

When induction motor is in no load, it draws a current ($I_o$) from the supply to produce flux in air gap and supply the iron loss. This current $I_o$ has two components namely $I_c$ and $I_m$, given in equation 4.16.

Where,

$I_c$ is active component which supplies no load losses

$I_m$ is magnetizing component creates the flux in core and air gap

$$\overline{I_o} = \overline{I_c} + \overline{I_m} \tag{4.16}$$

The equivalent circuit diagram of an induction motor is shown in Figure 4.15.

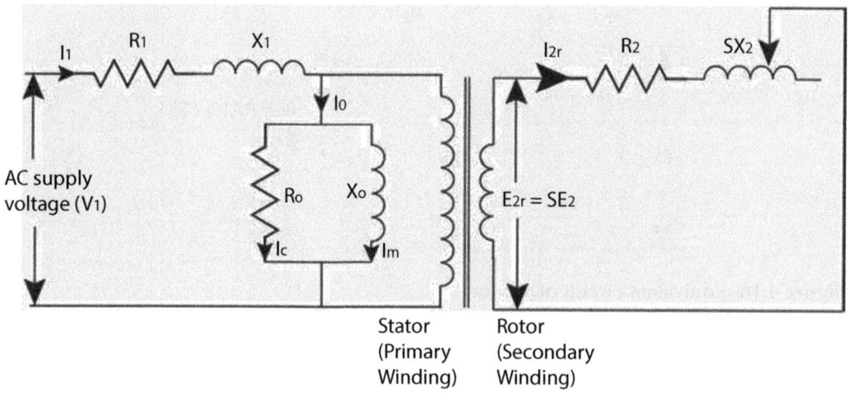


**Figure 4.15** Equivalent circuit diagram of an induction motor.

$$I_{2r} = \frac{E_{2r}}{X_{2r}} = \frac{S.E_2}{\sqrt{R_2^2 + (SX_2)^2}}$$

$$I_{2r} = \frac{E_2}{\sqrt{\left(\frac{R_2}{S}\right)^2 + X_2^2}} \tag{4.17}$$

$$\frac{R_2}{S} = R_2 + \frac{R_2}{S} - R_2$$

$$\therefore \frac{R_2}{S} = R_2 + R_2\left(\frac{1}{S} - 1\right) = R_2 + R_2\left(\frac{1-S}{S}\right)$$

So, the rotor resistance $\frac{R_2}{S}$ has two parts

1. Rotor resistance ($R_2$) represent copper loss.
2. Load resistance ($R_L$) = $R_2\left(\frac{1-S}{S}\right)$.

The equivalent circuit of a rotor is shown in Figure 4.16.
Equivalent circuit referred to Stator:
Transfer all the rotor parameters to stator side,

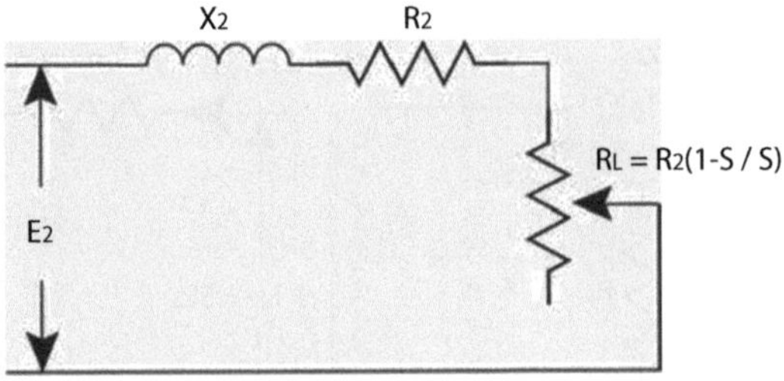


**Figure 4.16** Equivalent circuit of a rotor.

$k = \frac{E_2}{E_1}$ is transformation ratio

$$E_2' = \frac{E_2}{K}$$

$$I_{2r}' = KI_{2r} = \frac{K.SE_2}{\sqrt{R_2^2 + (SX_2)^2}}$$

$$X_2' = \frac{X_2}{K^2}$$

$$R_2' = \frac{R_2}{K^2}$$

$$R_L' = \frac{R_L}{K^2} = \frac{R_2}{K^2}\left(\frac{1-S}{S}\right) = R_2'\left(\frac{1-S}{S}\right) \tag{4.18}$$

The equivalent circuit is shown in Figure 4.17.
$R_{1e}$ is equivalent resistance referred to stator = $R_1 + R_2'$
$X_{1e}$ is equivalent reactance referred to stator = $X_1 + X_2'$

$$R_{1e} = R_1 + \frac{R_2}{K^2} \tag{4.19}$$

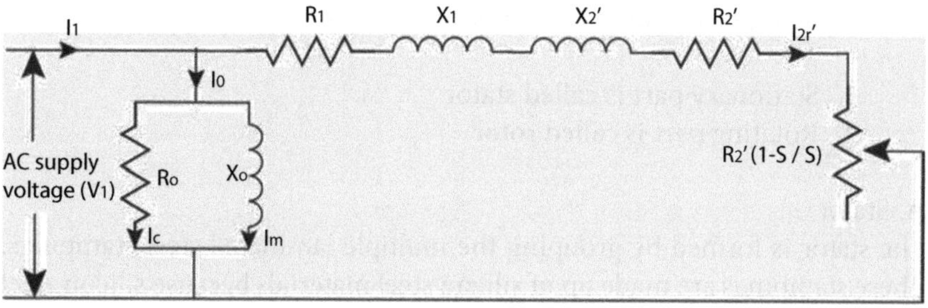


**Figure 4.17** Equivalent circuit.

$$X_{1e} = X_1 + \frac{X_2}{K^2} \tag{4.20}$$

While,

$\overline{I_o} = \overline{I_c} + \overline{I_m}$ and $\overline{I_1} = \overline{I_o} + \overline{I'_{2r}}$

The equivalent circuit is shown in Figure 4.18.

## 4.3 Single-Phase Induction Motor

### 4.3.1 Introduction

Single-phase induction motors are the low horsepower motors. These motors are used in lower horsepower applications like toys, fans, hand tools, hair dryers, etc [5]. The single-phase induction motor has a very simple construction.

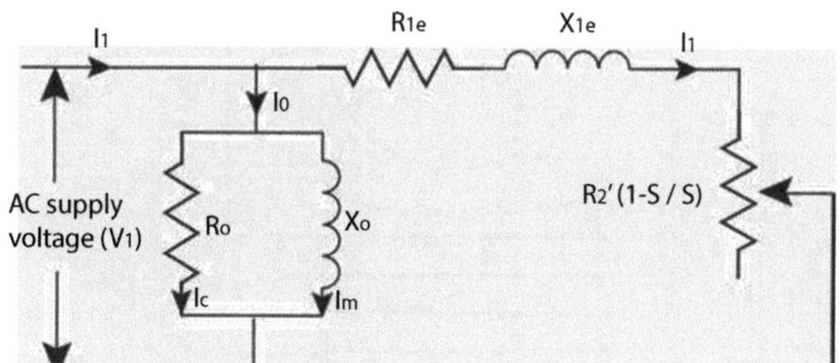


**Figure 4.18** Equivalent circuit.

The single-phase induction motor has two parts:

1. Stationary part is called stator
2. Rotating part is called rotor

A. Stator

The stator is formed by grouping the multiple laminated steel stampings. These stampings are made up of silicon steel materials because silicon steel has good permeability which is used to minimize the hysteresis loss. The lamination of the silicon steel is used to minimize the eddy current loss on it. The stator winding has definite number of poles and decided based on synchronous speed. The expression for synchronous speed is given in equation 4.21.

$$N_S = \frac{120 \times F}{P} \text{ in RPM} \tag{4.21}$$

The construction of the single-phase induction motor is shown in Figure 4.19.

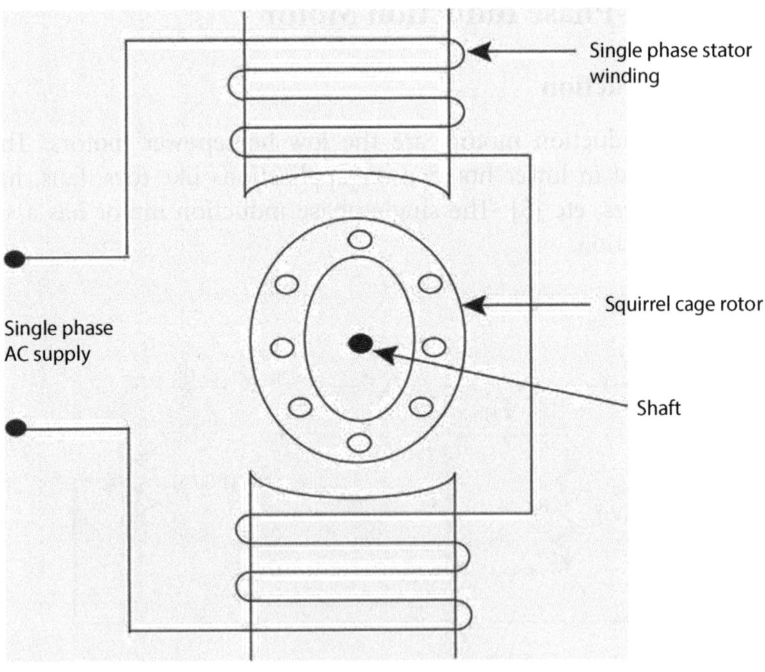


**Figure 4.19** Construction of the single-phase induction motor.

B. Rotor

The rotor of a single-phase induction motor is a squirrel cage type by construction. The construction of a squirrel cage rotor is shown in Figure 4.20. The rotor has squirrel cage bars permanently shorted to each other and the resistance of the rotor is very small. The main feature of this rotor is that it automatically adjusts itself for the same number of poles as that of the stator winding.

### 4.3.2 Working Principle

For motoring action, there should be two fluxes which interact with each other to produce the torque. Single-phase AC supply given to the stator winding and it produces the alternating flux. This flux is called main flux. This flux links with the rotor conductors and due to transformer action, EMF gets induced in the rotor. The induced EMF in the rotor drives the current since the rotor has the closed circuit. Due to this, rotor current produces the flux, which is called rotor flux. This single-phase induction motor is not a self-starting motor.

### 4.3.3 Types of Single-Phase Induction Motor

As a single-phase induction motor is not a self-starting, in order to make this motor self-starting, additional winding is added into the main stator winding. This winding is called as auxiliary winding. In order to produce magnetic field, at least two alternating fluxes are required with the phase difference. The main stator winding and auxiliary winding is electrically displaced by 90°. Now the single-phase induction motor starts as the two-phase induction motor.

The single-phase induction motor is classified into four types:

1. Split-phase induction motor

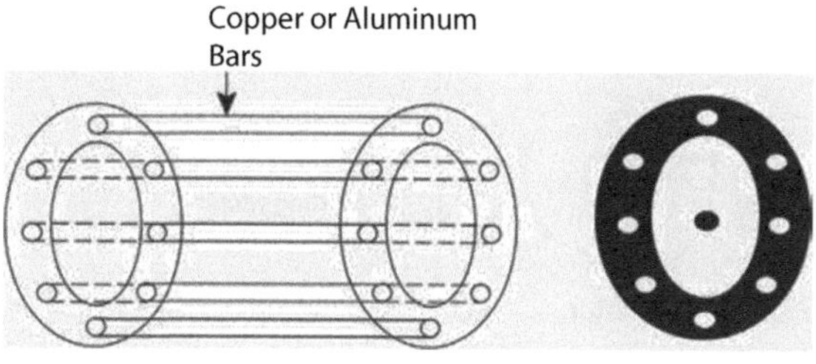


**Figure 4.20** Construction of rotor.

2. Capacitor-start induction motor
3. Capacitor-start and capacitor-run induction motor
4. Shaded-pole induction motor

**A. Split-Phase Induction Motor**

The schematic diagram of a split-phase motor is shown in Figure 4.21.

It has two stator windings, namely main winding or running winding and starting winding or auxiliary winding. These two running winding and starting winding are placed at a 90° angle electrically. The resistance and reactance of starting and running windings are selected to create the phase difference between these two windings. The auxiliary winding offers low reactance and high resistance; the main winding has high reactance and low resistance. When the speed reaches 75 to 80% of the synchronous speed, the centrifugal switch disconnects the auxiliary winding from the power supply. Only running winding alone is present in the circuit for the remaining operation time.

Let,

$I_m$ is current through main winding

$I_{st}$ is current through starting winding

The phase difference (α) between these two currents produces the flux as shown in Figure 4.22.

The speed vs. full load torque of a split-phase motor is shown in Figure 4.23.

**Applications**:

The applications of slip-phase induction motors in low starting current and moderate starting torque are as follows:

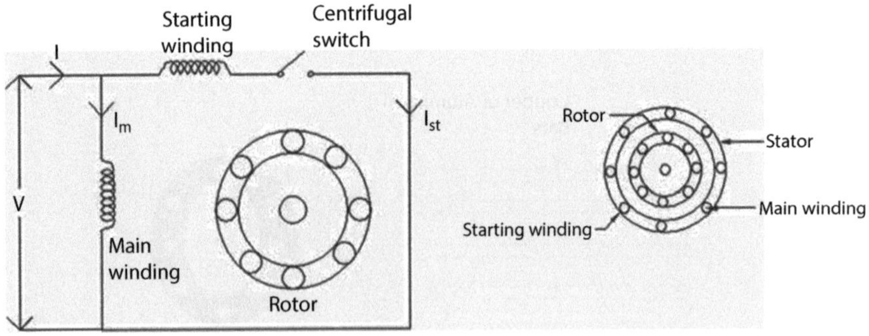


**Figure 4.21** Split-phase induction motor.

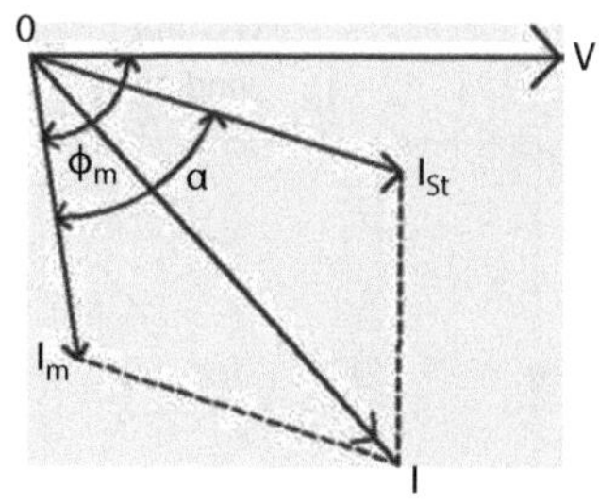


**Figure 4.22** The phase difference (α) between two currents.

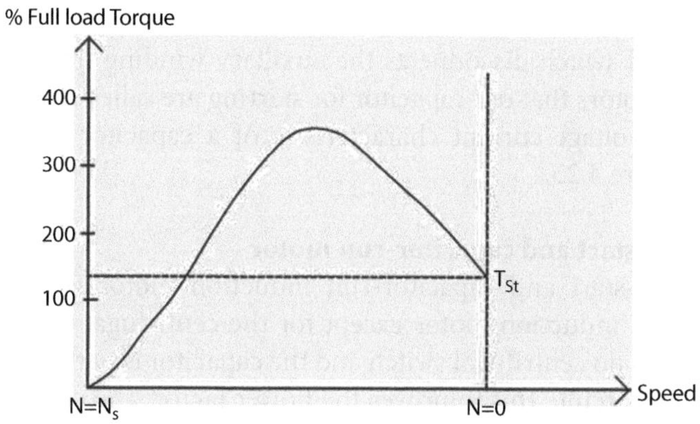


**Figure 4.23** Speed vs. full load torque.

i) Fans
ii) Blowers
iii) Grinders
iv) Centrifugal pumps
v) Washing Machines, etc.

**B. Capacitor Start Induction Motor**

A capacitor-start induction motor is similar to a split-phase type induction motor. The circuit diagram of a capacitor-start induction motor is shown in Figure 4.24.

The capacitor (C) is connected series with the starting (auxiliary) winding which draws leading current. The current $I_m$ lags the voltage by angle ($\varphi_m$), current $I_{st}$ leads the voltage by angle ($\varphi_{st}$). Starting torque developed by the motor is proportional to 'α' and it produce very high starting torque. When the speed reaches 75 to 80% of the synchronous speed,

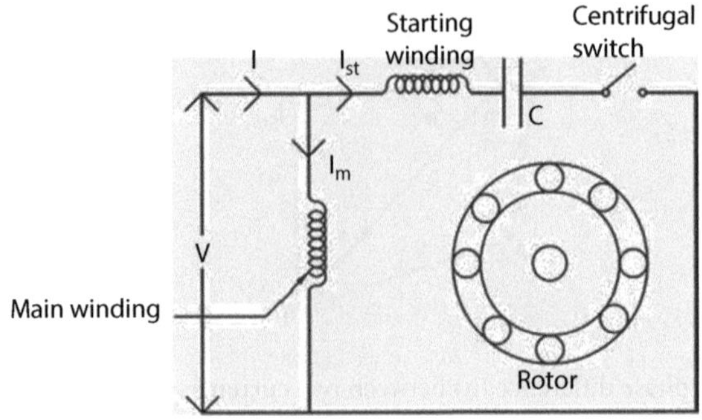


**Figure 4.24** Circuit diagram of capacitor-start induction motor.

the centrifugal switch disconnects the auxiliary winding from the power supply. The motors that use capacitor for starting are called capacitor-start motors. The voltage-current characteristic of a capacitor star motor is shown in Figure 4.25.

**C. Capacitor-start and capacitor-run motor**

The capacitor-start and capacitor-run induction motor is similar to a capacitor-start induction motor except for the centrifugal switch. In this motor, there is no centrifugal switch and the capacitor is permanently connected in the circuit. This improves the power factor. The circuit diagram of a capacitor-start and capacitor-run induction motor is shown in Figure 4.26.

The purpose of the capacitor is not only to start the motor; it is also required during the running time. So, the value of the capacitor 'C' is

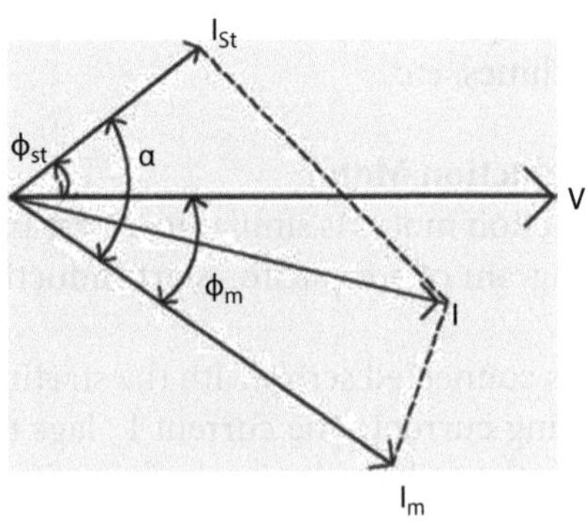


**Figure 4.25** Voltage and current characteristics.

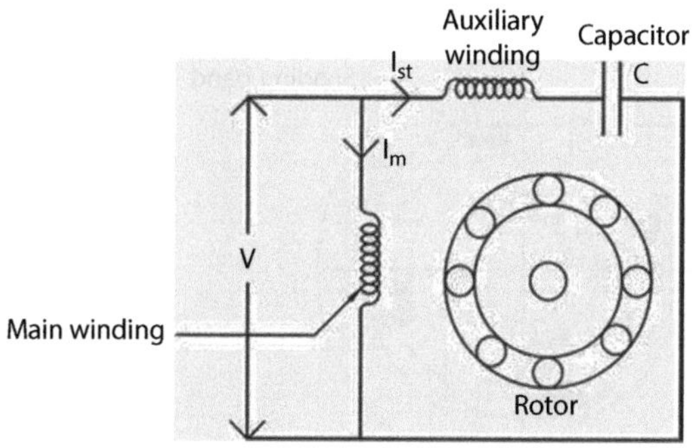


**Figure 4.26** Circuit diagram of capacitor-start and capacitor-run induction motor.

selected to combined requirements of both starting and best running condition. The speed vs. full load torque is shown in Figure 4.27.

**Application:**
The applications of capacitor-start and capacitor-run induction motors are as follows:

i). Compressor
ii). Air conditioning Machine

**D. Shaded-Pole Induction Motor**
The construction of a shaded-pole motor is shown in Figure 4.28.

The stator consists of a salient pole or projecting pole and the rotor consists of a squirrel cage. The stator poles are shaded, i.e., each pole carries a

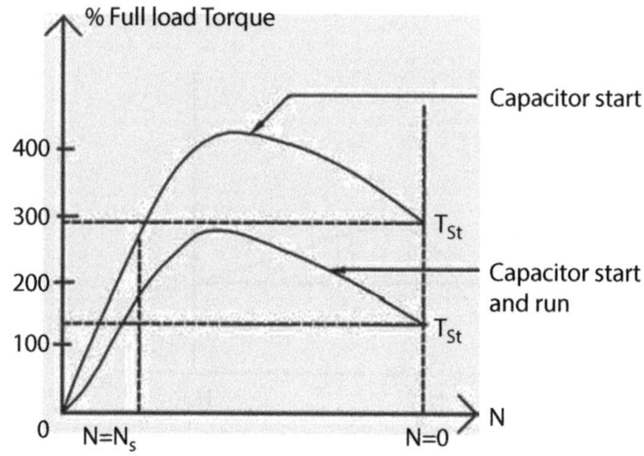


**Figure 4.27** Speed vs. full load torque.

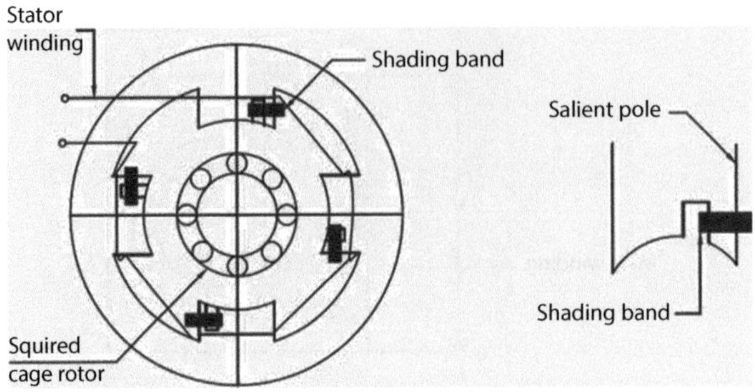


**Figure 4.28** Construction of shaded-pole motor.

copper band on one of its unequally divided part called shading band. The current carried by the stator winding is alternating and produces alternating flux. This flux will set up the currents in shading band. The speed vs. full load torque is shown in Figure 4.29.

**Applications:**

The applications of shaded-pole induction motors are as follows:

i) Small Fans
ii) Toy Motors
iii) Advertising Displays
iv) Film Projectors
v) Record Players, etc.

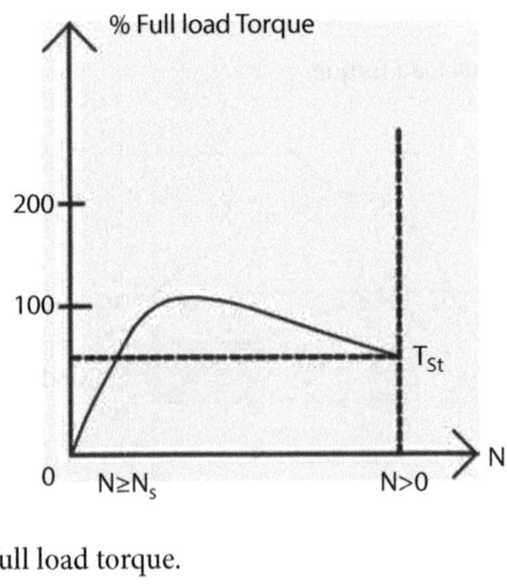


**Figure 4.29** Speed vs. full load torque.

## 4.4 Starting Methods of Induction Motor

### 4.4.1 Need for Starters

When a three-phase Induction motor is switched ON or powered ON at nominal supply voltage, due to zero back EMF at standstill condition, it draws the higher amount of current from the power supply [7, 14]. That is, when the induction motor is switched ON directly, it takes five to seven times its rated full load current as inrush current. Due to this excessive inrush current drawn from the supply, a severe voltage dip occurs at the time of motor starting [1]. This reduction of supply voltage is called voltage sag or dip. Due to the voltage sag or dip, other equipments connected to the same power supply are affected [8, 9]. In order to reduce the higher starting current drawn by the induction motor during starting time, starters are used. Starters are current limiting devices and can limit the current during motor starting time.

### 4.4.2 Types of Starters

Starters are generally classified into five types listed as follows:

- i) Direct On Line (DOL) starter
- ii) Primary resistor or reactor starter
- iii) Auto transformer starter
- iv) Star – Delta starter
- v) Rotor resistance starter

**A. Direct On Line starter**

A motor of small capacity less than 5 kW and which gathers normal speed quickly can be started with this starter. Figure 4.30 shows the three-phase induction motor with a DOL starter.

This starter has three main contactors ($M_1$, $M_2$ and $M_3$) Normally Open (NO) type, making and breaking the motor with power supply. These contactors are operated by a relay coil. The contactors ($S_2$, $S_3$ and $S_4$) which are Normally Closed (NC) type and are connected series with relay coil. The Over Load Release Coil (OLRC) is connected in series with motor line supply. The typical practical DOL starter is shown in Figure 4.31.

When TPST switch is closed, the Under Voltage Relay Coil (UVRC) is energized and it can operate the main contactors ($M_1$, $M_2$ and $M_3$) to close,

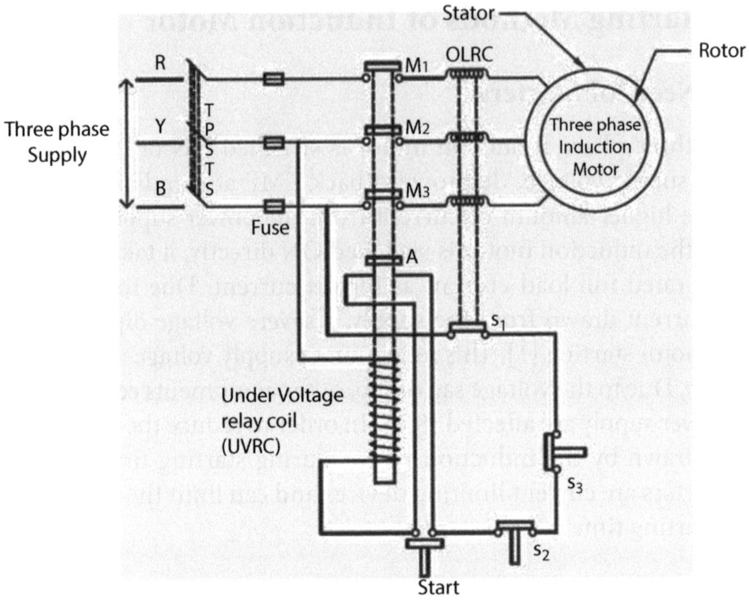


**Figure 4.30** Three-phase induction motor with DOL starter.

**Figure 4.31** Practical DOL starter (Courtesy: Larson and Toubro Limited).

full voltage supplied to the motor and motor starts rotates. Contactor $S_2$ is used to disconnect the supply to the motor by manually pressing it, and remote operation of the same can be achieved by contactor $S_3$.

No Voltage Protection:
Whenever the supply voltage is reduced below the specified value (under voltage) or failure of supply voltage, UVRC can release the main contactor ($M_1$, $M_2$ and $M_3$) and circuit becomes open circuit. Hence, the motor is protected from under voltage or loss of supply voltage.

Over Load Protection:
Whenever the current flow in the line exceeds the preset value, OLRC is energized and contactor $S_1$ becomes in the open position from the closed position. Once contactor $S_1$ becomes open, the UVRC is disconnected from the supply and it will release the main contactors, i.e., main contactors $M_1$, $M_2$ and $M_3$ becomes in an open position.

**B. Primary Resistor (or) Reactor Starter**

The circuit diagram of primary resistor or reactor is shown in Figure 4.32.

As the name implies, a variable resistor (or) reactor is connected in series between the power supply and stator (primary) terminals of the motor. The purpose of these resistors or reactors is to reduce the supply voltage, and this reduced voltage is applied to the motor terminals. This reduced voltage limits the current drawn by the motor during the time of starting. If the starting voltage across the stator terminals is reduced by 50%, then the starting current (inrush current) is reduced by 50% and torque is reduced to 25% of the full load voltage.

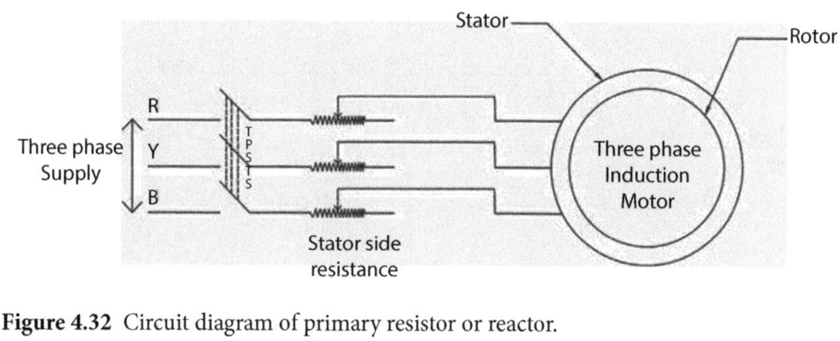


**Figure 4.32** Circuit diagram of primary resistor or reactor.

The advantages of this method are as follows:

i) Smooth Acceleration
ii) High Power factor during starting
iii) Less expensive
iv) Closed transition Starting

The disadvantages of this method are as follows:

i) Power Lost in Resistors
ii) Low starting torque
iii) Less efficiency

**C. Auto Transformer Starter**

Figure 4.33 shows the circuit diagram of autotransformer-based starting method of the three-phase induction motor.

This starter is used to reduce input voltage supplied to the three-phase induction motor to limit the starting current (inrush current) at the time of starting. The reduced voltage is obtained by an auto transformer. When the handle is moved to the start position, the motor is connected through the

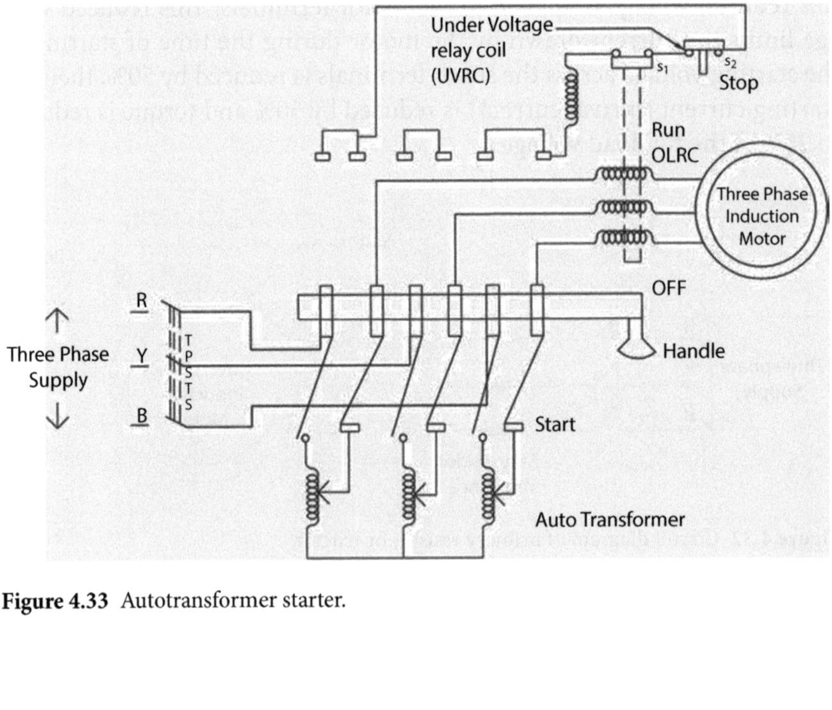


**Figure 4.33** Autotransformer starter.

auto transformer. By varying the autotransformer, input supply voltage is reduced and the reduced voltage is applied to stator terminals. Hence starting current is reduced. Once the motor reaches 80% of the nominal speed, the handle moves to the RUN position and the autotransformer is bypassed. In this position, full voltage is applied to the stator terminals and motor working its rating. The line current drawn by the motor from the supply is much less in this type as compared to primary resistance starter.

Overload Protection:
Whenever the motor draws current exceeding the preset value due to overload, the OLRC is energized and operates the contactor $S_1$, hence the power supply to the motor is disconnected.

Low Voltage Protection:
The Under Voltage Relay Coil (UVRC) is connected across any two lines. Whenever supply voltage goes below the preset value (under voltage) or loss of power supply, UVRC de-energizes and releases the handle to OFF position.

The advantages of this method are as follows:

i) Reduced Line Current
ii) Smooth Starting
iii) High acceleration
iv) Less power loss during starting due to inductive coil

The disadvantages of this method are as follows:

i) Cost is high
ii) It is economically suitable for large motor only

**D. Star-Delta Starter**

The circuit diagram of a star-delta starter is shown in Figure 4.34. As the name implies, this method has both star and delta connection type for starting and running. It consists of a two-way switch (one position is start and one position is run) which converts the star position (starting) to delta position (running).

When the two-way switch is at START position, the stator windings are connected in star. Therefore the applied voltage is reduced by a factor of

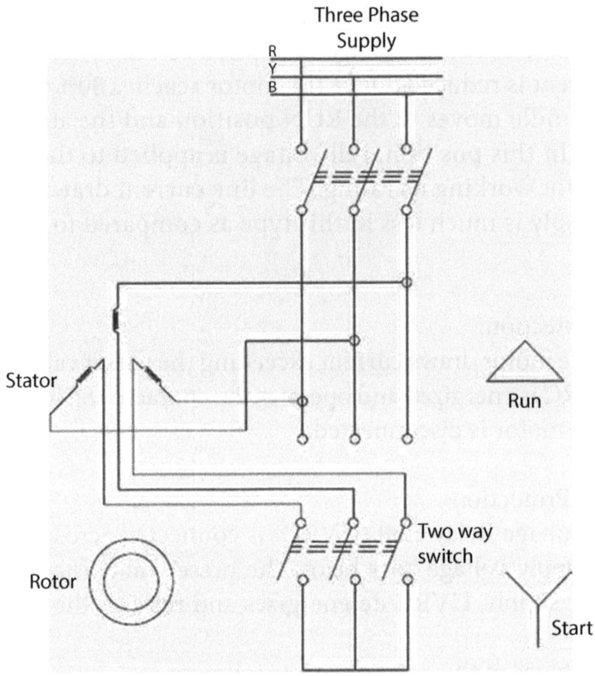


**Figure 4.34** Star-Delta starter.

$\frac{1}{\sqrt{3}}$ hence the starting current is reduced. When the motor reaches 70% to 80% of nominal speed, the switch is changed to RUN position and the motor is connected in delta. Therefore full voltage is applied to the motor in this running condition. A Star-Delta starter is much cheaper compared to an auto-transformer starter. This starter is mainly used for small and medium-size motors.

### E. Rotor Resistance Starter

This starter is used for slip ring induction motor only. The circuit diagram of a slip ring induction motor started by a rotor resistance starter is shown in Figure 4.35.

In this method, external resistance or starting resistance is connected to the rotor terminals. The motor is always started with full line voltage applied across the stator terminals and the starting current is controlled by introducing the variable resistance in the rotor circuit. At starting time, the full resistance added into the rotor circuit and hence the starting current

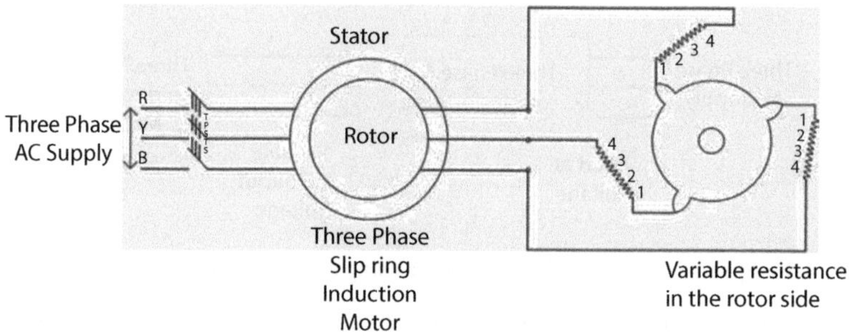


**Figure 4.35** Slip ring induction motor.

is reduced. The resistance is gradually reduced in the rotor circuit as the motor gathers speed.

## 4.5 Speed Control of Three-Phase Induction Motor

The speed of an induction motor can be controlled by two main methods:

i) Stator Side Control
ii) Rotor Side Control

The first method is used for both squirrel cage and slip ring induction motors, while the second method is used for only slip ring induction motors.

**A. Stator Side Control**
**i. Stator voltage control**

The speed of the induction motor can be controlled by varying the applied voltage to the stator winding. This type of speed control is known as stator voltage control. In this method, the supply frequency is constant. The stator voltage can be controlled by two methods:

1. Using auto transformers
2. Primary resistors connected series with stator winding

a) Using Auto Transformers
The auto transformer based speed control of a three-phase induction motor is shown in Figure 4.36.

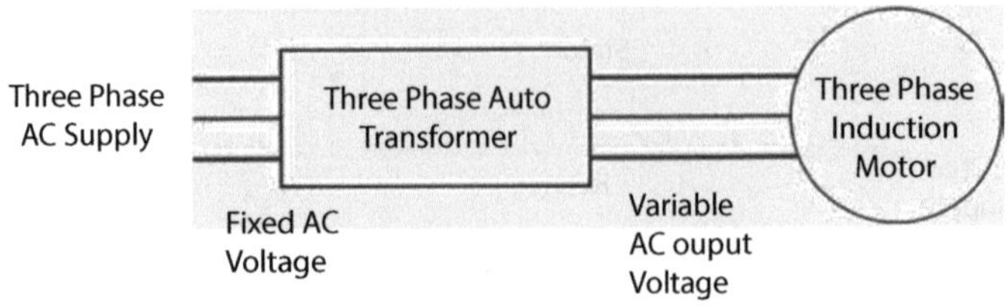


**Figure 4.36** Speed control of induction motor using auto transformer.

The input to the auto-transformer is fixed AC voltage supply. By varying the auto transformer, variable AC output voltage without change in frequency is achieved. This variable voltage is applied to the induction motor and the induction motor speed is controlled.

b) Primary Resistors connected in series with stator winding
The primary resistor based speed control of a three-phase induction motor is shown in Figure 4.37.

By varying the primary resistance the connected to the stator terminals reduces the voltage applied to the motor. Hence, the motor speed can be controlled. In this method, power loss occurs in the primary resistors are high. The torque is proportional to the square of the stator voltage (i.e., T α $V^2$) and by varying the voltage the torque also changes.

**ii. Stator frequency control**

The stator frequency control is one of the methods to control the speed of the three-phase induction motor. In this method, input supply frequency to the induction motor is varied. Under no load condition, the induction motor operates in the small-slip region and where the speed of the induction motor is close to the synchronous speed of the rotating magnetic field. The synchronous speed of the induction motor is given in equation 4.22.

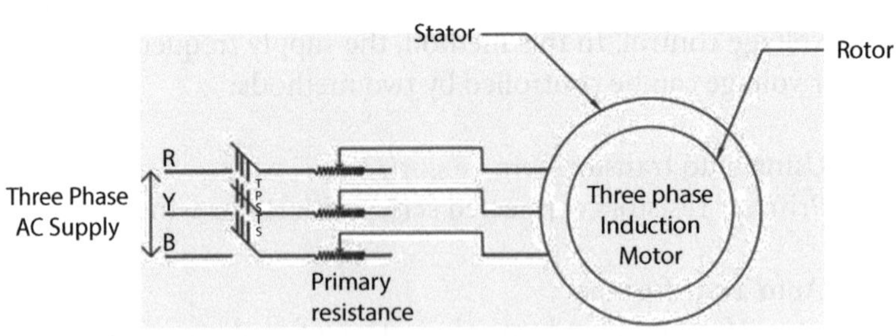


**Figure 4.37** Speed control of induction motor using primary resistor

$$N_s = \frac{120F}{P} \tag{4.22}$$

Figure 4.38 shows the schematic diagram of speed control based on stator frequency control.

From the equation 4.22, synchronous speed of the motor is directly proportional to the frequency of the supply voltage, i.e., change in the supply frequency has an impact on the motor speed. In order to control the motor speed, input supply frequency is varied. This method is rarely used. The air gap flux of induction motor is inversely proportional to the frequency and directly proportional to the voltage. If there is any decrease in the frequency then the airgap flux gets saturated. In order to avoid this, the Voltage/Frequency method is widely used for speed control of an induction motor.

**iii. Voltage/Frequency Control**

The schematic diagram of voltage (V)/frequency (F) control is shown in Figure 4.39.

The V/F control can be achieved by using power electronic convertors. The fixed AC voltage and frequency is fed to the rectifier circuit and it converts the AC into DC. This DC supply is fed to the invertor circuit and it converts DC into variable AC voltage and variable frequency. This controlled V/F is fed to the stator of the induction motor and speed of the

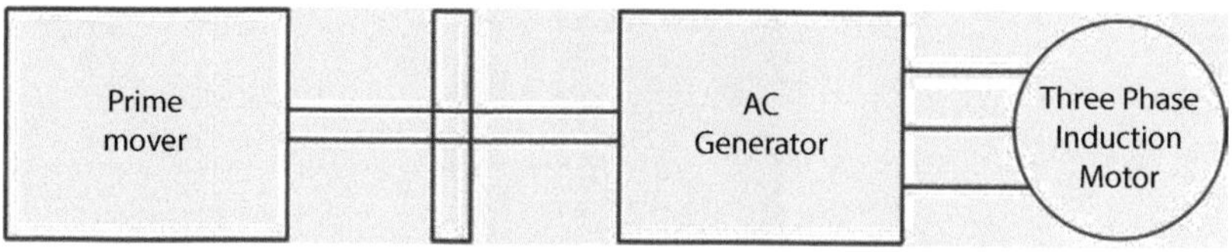


**Figure 4.38** Stator frequency control.

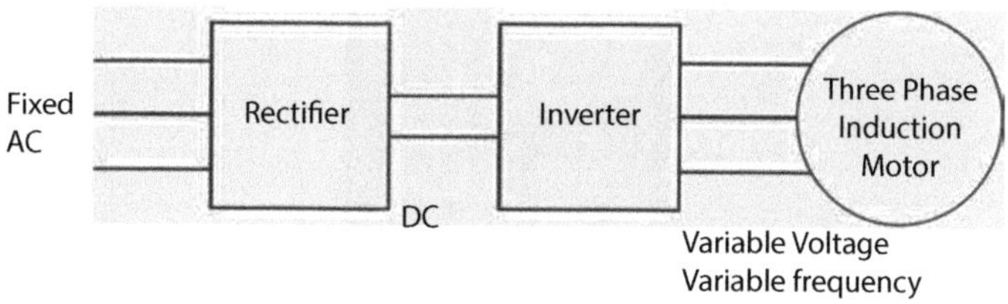


**Figure 4.39** Schematic diagram of V/F control.

induction motor is controlled. This method is a widely used method of induction motor speed control.

Figure 4.40 shows the typical V/F control used for speed control of the induction motor in air handling unit application.

**B. Rotor side control**

**i. Cascade Control**

The schematic diagram of cascaded type speed control of slip ring induction motor is shown in Figure 4.41.

This method consist of two slip ring induction motors, the first motor is called main motor $M_1$ and this motor is coupled with a second motor or auxiliary motor $M_2$. The three-phase supply is fed to the stator of the main motor $M_1$, the slip ring voltage of motor $M_1$ is fed to the stator of the auxiliary motor $M_2$. In this cascading method, if both motors produce the torque in the same direction it means it is cumulative cascading, and

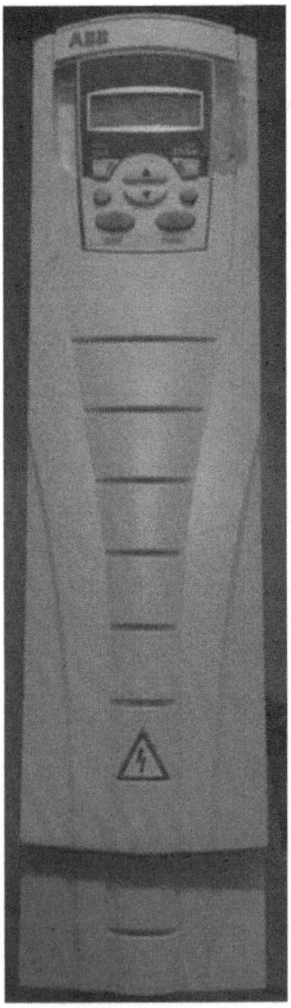

**Figure 4.40** Typical V/F controller. Courtesy: ABB

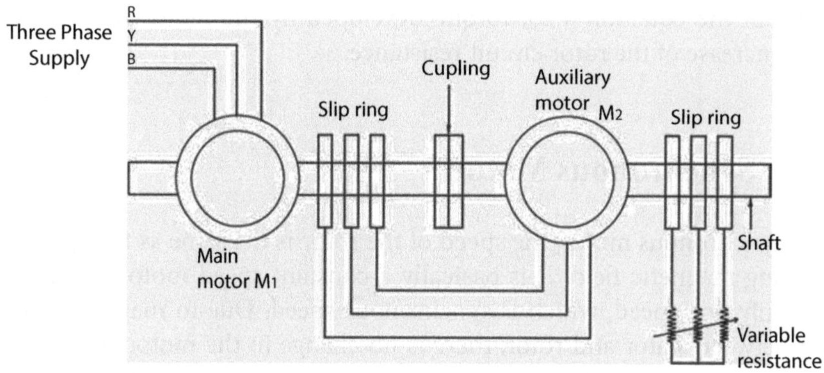


**Figure 4.41** Cascaded type speed control for slip ring induction motor.

opposite direction means it is differential cascading. This method is also known as Tandem control.

**ii. Adding external resistance in the Rotor circuit**

The circuit diagram of speed control of slip ring induction motor by external resistance is shown in Figure 4.42.

This method is applicable only for slip ring Induction motor. The external resistance is connected to the rotor circuit of slip ring induction motor. A simple and primitive method of speed control of a slip ring induction motor is achieved by varying the external rotor resistance.

The expression for starting torque developed by slip ring induction motor is given in equation 4.23.

$$T_{st} \alpha \frac{E_2^2 R_2}{R_2^2 + X_2^2} \tag{4.23}$$

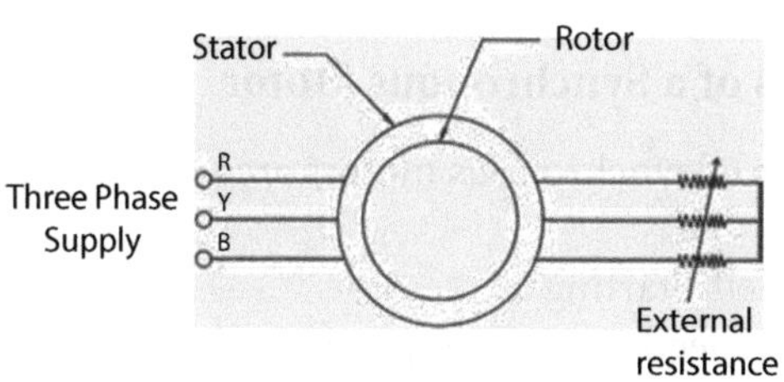


**Figure 4.42** Speed control of slip ring induction motor by external resistance.

From the equation 4.23, torque developed by the motor is increased with increase of the rotor circuit resistance.

## 4.6 Synchronous Motor

In a synchronous motor, the speed of the rotor is the same as the speed of rotating magnetic field. It is basically a constant speed motor because it has only one speed, which is synchronous speed. Due to magnetic locking between stator and rotor, there is no change in the motor speed with respect to change in the load. In other words, it's in synchronism with the supply frequency. The synchronism is always present between the stator and rotor up to 90° electrically [1]. The expression for speed of the synchronous motor is given in equation in 4.24.

$$\text{Synchronous speed, } N_s = \frac{120F}{P} \tag{4.24}$$

Where,

F is supply frequency in Hz

P is number of poles

### 4.6.1 Construction

The construction of a synchronous motor is similar to the three-phase induction motor, except the rotor is excited by external DC supply, the reason of which is explained later. Three-phase AC power supply is given to the stator winding and DC supply is given to rotor winding for their operation in synchronous motor. The construction of the synchronous motor is shown in Figure 4.43.

### 4.6.2 Features of a Synchronous Motor

The main features of synchronous motors are as follows:

i) It is not Self-Starting
ii) Constant speed motor irrespective of load
iii) Used for power factor enhancement

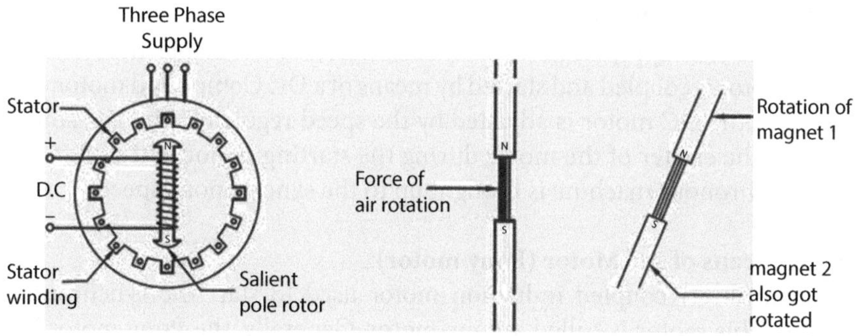


**Figure 4.43** Construction of synchronous motor.

### 4.6.3 Working Principle

A synchronous motor needs two separate electric power supplies as an input because it is a doubly excited motor. The stator has the three-phase armature winding and is powered by three-phase AC power supply, while the rotor has field winding on it and is powered by DC power supply. The current flow in the three-phase stator winding produces the rotating magnetic flux and the DC current flow in the rotor produces the constant flux. For 2 pole machine at 50 Hz frequency, the rotating magnetic field produced by the armature winding rotates 3000 revolutions/min or 50 revolution/sec. At a particular instant, rotor and stator poles might be the same polarity (N-N or S-S), causing a repulsive force on the rotor, and the very next instant it will be N-S, causing attractive or repulsive due to inertia of the rotor. Due to this, the rotor is unable to rotate in any direction and remains in a standstill condition. Hence, this motor is not a self-starting motor. In order to make the rotation, the rotor is initially coupled with external devices and it rotates the rotor in the same direction as the rotating magnetic field. Magnetic locking occurs and the rotor of the synchronous motor rotates in the speed of the rotating magnetic field.

### 4.6.4 Starting Methods of Synchronous Motor

The synchronous motor is not a self-starting motor, so in order to start the motor any of the following methods are used:

i. By means of DC motor
ii. By means of AC motor or Pony motor
iii. By means of damper grids in the pole faces

**A. By means of DC motor**

Wherever DC supply and DC compound motor are available, the synchronous motor is coupled and started by means of a DC Compound motor. The speed of the DC motor is adjusted by the speed regulator. The DC source powers the exciter of the motor during the starting period, then after that the synchronous machine is brought up to the synchronous speed.

**B. By means of AC Motor (Pony motor)**

A small direct coupled induction motor used to start the synchronous motor. This motor is called a Pony motor. Generally, the Pony motor has two poles less than the synchronous motor, so that it is capable of raising the speed up to the synchronous speed. After normal operation is established, the pony motor remains connected with the synchronous motor or uncoupled from the synchronous motor. This method is not very sufficient (satisfactory) and not suitable for industrial needs.

**C. By means of Damper Grids in the Pole Faces**

In order to make the synchronous motor self-starting, special winding is provided on the rotor poles called damper winding or squirrel cage winding. The damper winding consists of short circuited copper bars embedded in the face of field poles. Whenever the AC supply is given to the stator winding of the synchronous motor, it produces the rotating magnetic field, which makes the rotor rotate like a squirrel cage induction motor. The exciter moves along the rotor; when the motor attains about 95% of synchronous speed, the rotor winding is connected to exciter terminals and rotor is magnetically locked by the rotating magnetic field produced by the stator and the motor runs as a synchronous motor.

### 4.6.5 Torque Equation of Synchronous Motor

The mechanical power developed per phase in the rotor is given in equation 4.25.

$$P_{mech} = E_b I \text{Cos}\phi \tag{4.25}$$

$$\frac{P_{mech}}{Phase} = \frac{E_b V}{Z_s}\text{Cos}(\theta - \delta) - \frac{E_b^2}{Z_s}\text{Cos}\theta \tag{4.26}$$

Max power developed by motor is given in equation 4.27.

$$(P_{mech})_{Max} = \frac{E_b V}{Z_s} - \frac{E_b^2}{Z_s} \text{Cos}\theta \tag{4.27}$$

$$\therefore \text{Cos } \theta = 0$$

$$P_{mech} = \frac{E_b V}{X_s} \text{Cos}(90 - \delta) = \frac{E_b V}{X_s} \text{Sin}\delta \tag{4.28}$$

$$(P_{mech})_{Max} = \frac{E_b V}{Z_s}; \text{ when } \delta = 90 \tag{4.29}$$

The various torque associated with a synchronous motor are

- Starting Torque
- Running Torque
- Pull in Torque
- Pull out Torque

## 4.7 Stepper Motor

Stepper motors are also called stepping or step motors. The name stepper is used because this motor rotates through a fixed angular step in response to each input current pulse received by its controller. The unique feature of a stepper motor is that output shaft rotates in a series of discrete angular intervals or steps taken when each command pulse is received. When a definite number of pulses are supplied, the shaft turns through a definite known angle [16].

Step angle:

The step angle ($\beta$) is the angle in which the stepper motor shaft rotates based on the each command pulse. If the step angle is small, the greater the number of steps per revolution and the higher the resolution or accuracy of positioning obtained. The stepper motor step angle is shown in Figure 4.44.

The step angles can be as small as 0.72° or as large as 90°. The most common step sizes are 1.8°, 2.5°, 7.5° and 15°.

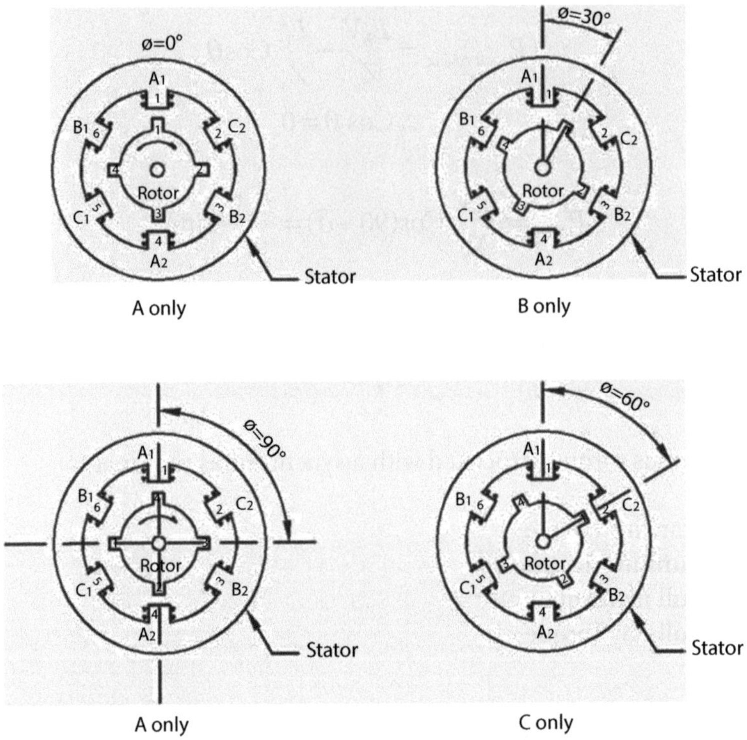


**Figure 4.44** Step angle of stepper motor.

i) The formula for step angle ($\beta$)

$$\beta = \frac{N_s - N_r}{N_s N_r} \times 360° \tag{4.30}$$

Or

ii) Step angle

$$\beta = \frac{360°}{m \times N_r} \tag{4.31}$$

Where,

m is number of stator phases

$N_S$ is number of stator poles

$N_r$ is number of rotor poles

Resolution:

It is defined as the number of steps needed to complete one revolution of the rotor shaft.

$$\therefore \text{Resolution} = \frac{\text{No of steps}}{\text{Revolution}} = \frac{360^\circ}{\beta} \tag{4.32}$$

The stepper motor is classified into following types

i) Variable Reluctance (VR) motor
ii) Permanent Magnet (PM) stepper motor
iii) Hybrid stepper motor

The stepper motors are used in following applications,

i) Instrumentation application
ii) Computer peripherals
iii) Office equipment
iv) Machine tool application
v) Robotics

## 4.8 Brushless DC (BLDC) Motor

A BLDC motor is a permanent magnet synchronous motor which is driven by DC power supply.

### 4.8.1 Construction

The construction of a BLDC motor is shown in Figure 4.45. Depending on the stator windings, this motor can be configured as single-phase, two-phase or three phase- motors. The stator of a BLDC motor is made up of stacked steel laminations to carry the windings, which are placed in a slot; the slots are axially cuts along the inner periphery of the stator. Each winding is constructed with several interconnected coils, where one or more coils are placed in each slot. In order to form the even number of poles, each of these windings is distributed over the stator periphery.

A BLDC motor has a permanent magnet in the rotor. The number of poles in the rotor varies from 2 to 8 poles pairs with alternate south and north poles depending on the application requirements.

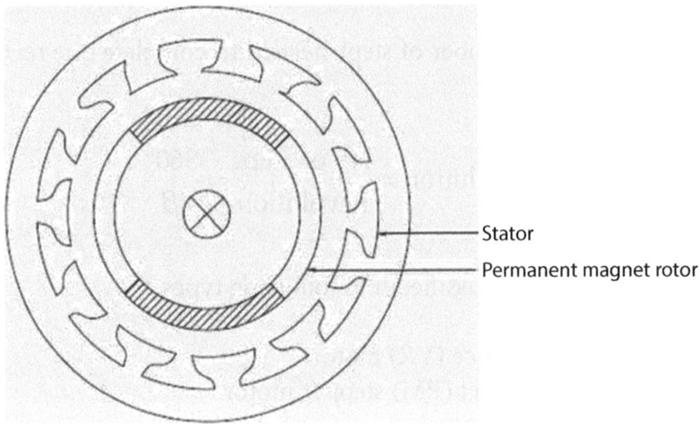


**Figure 4.45** Construction of BLDC motor.

### 4.8.2 Working Principle

When the stator windings are powered by the electric power supply, it becomes the electromagnet and starts producing the uniform magnetic field in the air gap. When DC supply is given, switching arrangement makes to generate an AC voltages waveform with trapezoidal shape. The rotor continues to rotate due to the force of interaction between electromagnet (stator) and permanent magnet rotor.

The BLDC motors are used in the following applications,

i) Computer Hard drives and DVD/VCD players
ii) Electric Vehicles, Hybrid Vehicles
iii) Washing Machines, Compressors and dryers
iv) Fans, pumps and Blowers

## 4.9 Alternator

An alternator is the electromechanical device or equipment which is used to convert the mechanical energy into electrical energy [10–12]. Alternators are similar to generators in DC machine, and the output of an alternator is AC supply, unlike generator output, which is DC supply.

### 4.9.1 Construction

The alternator has two main parts, namely stationary part (stator) and rotating part (rotor).

Stationary Winding → Stator → Armature
Rotating Winding → Rotor → Field

**A. Stator**

The stator is a stationary part of the alternator. It consists of stator core, slots and armature winding. Armature winding is placed on the slots similar to DC generator field windings. The stator core is made up of laminated steel stampings insulated from each other with varnish or paper. This laminated steel stamping is used to reduce the eddy current losses. The arrangement of armature winding in the slot is shown in Figure 4.46.

**B. Rotor**

The rotor is a rotating part of the alternator [13]. It carries a field winding wound on it. Based on the construction of the rotor, it is classified into two types:

i. Salient pole or projecting pole
ii. Smooth cylindrical type

**i. Salient Pole rotor or projecting pole**

As the name implies, the poles in the rotors are projected out from the surface. So, salient pole rotor is also called projecting pole rotor. These poles are made up of laminated steel stampings. The construction of the salient type rotor is shown in Figure 4.47.

The pole face has been given a specific shape as shown in Figure 4.47. This type of rotor has small axial length and a larger diameter. As shown in Figure 4.47, field windings are wound on the pole shoe. The centrifugal

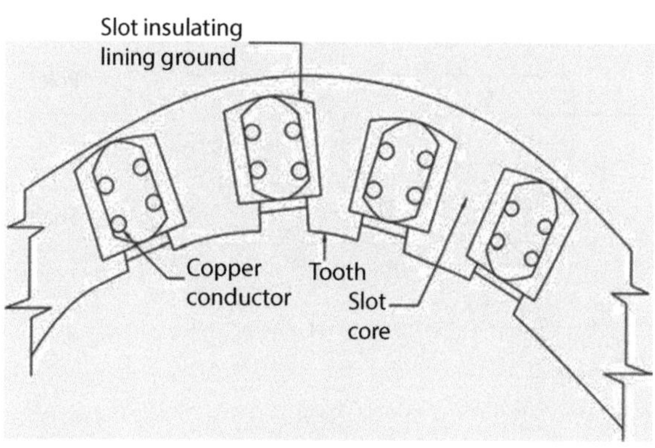


**Figure 4.46** Arrangement of armature winding in the slot.

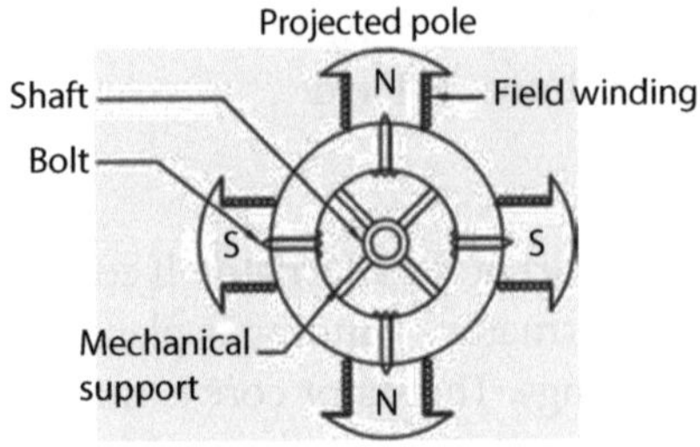


**Figure 4.47** Salient pole type rotor.

force acting on the rotor is the limiting factor of rotor size. Salient pole rotors are used for low speed application ranging from 125 RPM to 500 RPM. Generally, this type of rotor is used in hydro power and diesel power station driven by IC engines or water turbines.

**ii. Smooth Cylindrical Type**

As the name implies, this type of rotor's outer surface is smooth, unlike the salient pole type. The rotor consists of a smooth, solid steel cylinder having a number of slots to accommodate the field winding. The slots are covered at the top with the help of steel or manganese wedges. The construction of a smooth cylindrical type rotor is shown in Figure 4.48. The unslotted

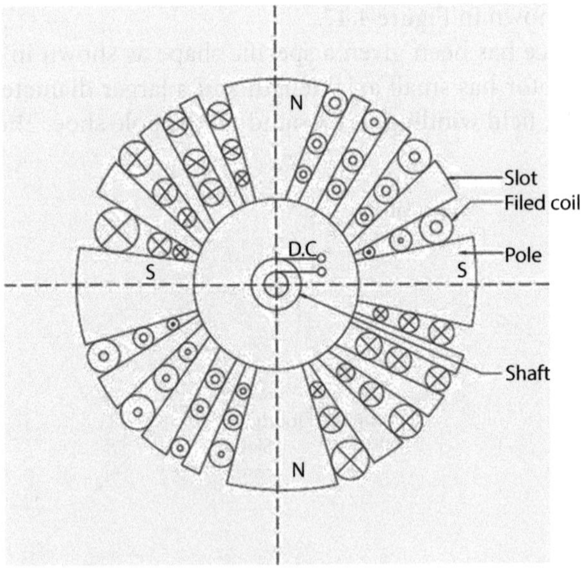


**Figure 4.48** Smooth cylindrical type rotor.

portions of the cylinder act as the poles. As the outer surface of the rotor is smooth, it maintains the uniform air gap between stator and rotor. This type of rotor has a small diameter and large axial length. The main advantage of this rotor is that it is mechanically strong, and it is used in high-speed application ranging from 1500 RPM to 3000 RPM. Generally, this type of rotor is used in thermal power plants driven by steam turbines.

The comparison of salient pole type and smooth cylindrical type rotors are listed in Table 4.1.

### 4.9.2 Working Principle

The alternators work on the principle of Faraday's law of electromagnetic induction. Whenever the conductor is placed in the magnetic, due to the relative motion between the conductors and the magnetic flux, EMF gets induced on it. The DC generator also works on the same principle; the basic difference is armature winding in the alternator is stationary while

**Table 4.1** Difference between salient and cylindrical type of rotor.

| S. No. | Salient pole rotor | Smooth cylindrical rotor |
|---|---|---|
| 1 | Air gap is non-uniform due to projecting surface | Air gap is uniform due to smooth cylindrical surface |
| 2 | The poles are projecting out from the surface | The unslotted portion of the cylinder acts as poles, hence poles are non-projecting |
| 3 | Diameter is high and small axial length | Diameter is small and large axial length |
| 4 | Mechanically weak | Mechanically robust |
| 5 | Applicable for low-speed alternator | Applicable for high-speed alternator |
| 6 | Driven by water turbine and IC engine | Driven by steam turbines |
| 7 | For same size, the rating is smaller as compared with cylindrical type | For same size, rating is higher as compared with salient pole type |
| 8 | Separate damper winding is provided | Separate damper winding is not necessary |

field winding in the DC generator is rotating. The relative motion of a single conductor under the magnetic field produced by two stationary poles is shown in Figure 4.49.

**Position 1:**
At position 1, the entire velocity of the component is parallel to the flux lines and there are no flux linkages by the conductor, So rate of change of frequency $\frac{d\phi}{dt} = 0$ and induced EMF also zero.

**Position 1-2:**
The part of the velocity component perpendicular to the flux lines and proportional to EMF induced on the conductor.

**Position 2:**
The velocity component exact perpendicular to the flux lines and EMF induced is maximum at its positive region.

**Position 2-3:**
The velocity component is perpendicular to the flux lines and flux starts decreasing and induced EMF also starts decreasing.

**Position 3:**
The conductor is again parallel to the flux lines and induced EMF on the conductor is zero.

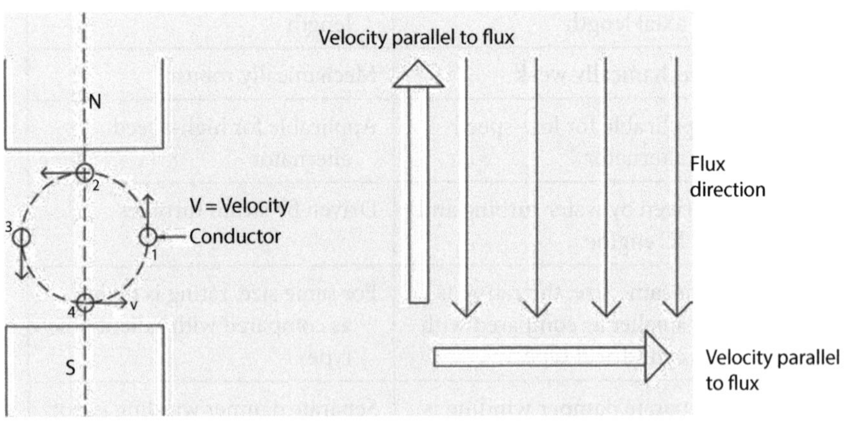


**Figure 4.49** Magnetic field.

**Position 3-4:**
The velocity component is perpendicular to the flux lines and travels in the opposite direction (negative).

**Position 4:**
The conductor is exactly perpendicular to the flux lines and the induced EMF is maximum in the opposite direction.

The flux linkage is shown in Figure 4.50.

Four pole alternator,

$$360^\circ \text{ mechanical} = 720^\circ \text{electrical}$$

$$360^\circ \text{ mechanical} = 360^\circ \times \frac{P}{2} \text{electrical}$$

Where,

P is number of poles

i.e., $1^\circ \text{Mechanical} = \left(\frac{P}{2}\right)^\circ$ electrical

Frequency of Induced EMF:

The frequency of the induced EMF is given in equation 4.33.

$$F = \frac{P}{2} \times \frac{N}{60} \tag{4.33}$$

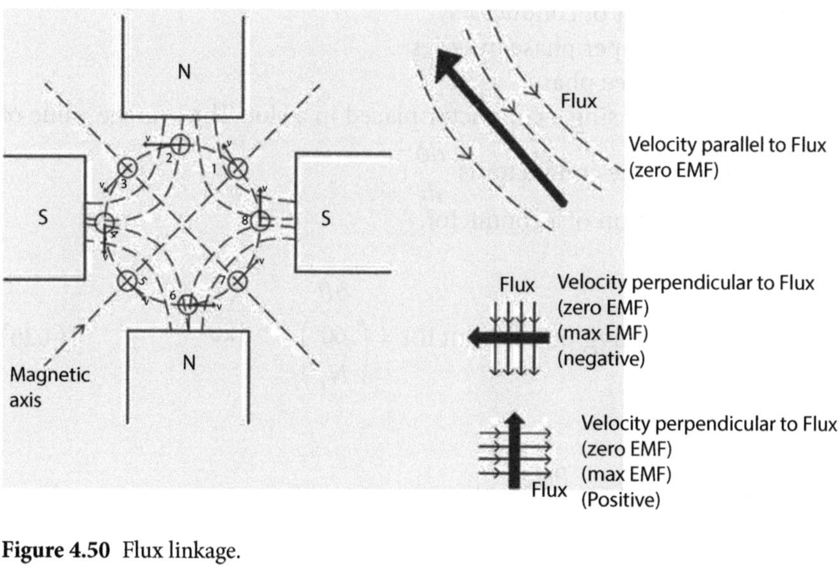


**Figure 4.50** Flux linkage.

$$F = \frac{PN}{120}\,\text{Hz (cycles per sec)} \tag{4.34}$$

Where,

One mechanical revolution of motor $= \frac{P}{2}$ cycles of EMF electrically

Synchronous Speed ($N_S$):

The expression for synchronous speed is given in equation 4.35.

$$N_s = \frac{120F}{P} \tag{4.35}$$

Where,

$N_S$ is synchronous speed (RPM)

F is frequency (Hz)

P is number of poles

### 4.9.3 EMF Equation of an Alternator

Let,

Φ is flux per pole in Wb

P is number of poles

$N_s$ is synchronous speed in RPM

F is frequency in Hz

Z is total number of conductors

$Z_{ph}$ is conductor per phase in series

$Z_{ph}$ = Z/3 for three phase

Considering the single conductor placed in a slot. The average value of EMF induced in one conductor is $\frac{d\phi}{dt}$.

For one revolution of a conductor,

$$\text{Eavg per conductor} = \frac{\phi P}{\left(\frac{60}{N_s}\right)} = \phi \frac{PN_S}{120} \tag{4.36}$$

But $F = \frac{PN_S}{120}$, hence $\frac{PN_S}{60} = 2F$

$E_{avg}$ per conductor = $2F\phi$ volts
EMF per turn = 2×(EMF per conductor)
= $2 \times (2F\phi)$
= $4F\phi$ Volts
Average $E_{ph}$ = $T_{ph}$ x (Average EMF per Turn)

$$= T_{ph} \times 4F\phi \quad (4.37)$$

∴ The Form Factor, $K_F = \frac{\text{RMS Value}}{\text{Average Value}} = 1.11$ for sinusoidal wave shape
RMS value of

$$E_{ph} = K_F \times \text{Average Value}$$

$$= 1.11 \times 4F\phi T_{ph}$$

$$E_{ph} = 4.44F\phi T_{ph} \text{ Volts} \quad (4.38)$$

Where,
Tph is number of turns per phase

$$T_{ph} = \frac{Z_{ph}}{2}$$

**Pitch Factor or Coil Span Factor ($K_C$):**

$$K_C = \frac{E_R \text{ when coil is Shor Pitched}}{E_R \text{ when coil is Full Pitched}}$$

$$= \frac{2E\cos\left(\frac{\alpha}{2}\right)}{2E} \quad (4.39)$$

$$K_C = \cos\left(\frac{\alpha}{2}\right)$$

α = Angle of short pitch

**Distribution Factor ($K_d$):**

$$K_d = \frac{\text{ER when coils are disconnected}}{\text{ER when coils are connected}} = \frac{2R\,\text{Sin}\left(\frac{m\beta}{2}\right)}{2m\,R\text{Sin}\left(\frac{\beta}{2}\right)}$$

$$K_d = \frac{\text{Sin}\left(\frac{m\beta}{2}\right)}{m\text{Sin}\left(\frac{\beta}{2}\right)}; \text{m} = \text{slots per pole per phase} \tag{4.40}$$

$\beta$ = slot angle $= \frac{180^\circ}{n}$ and n= slot per pole

Generalized EMF equation is given in equation 4.42.

$$E_{ph} = 4.44 \times K_C K_D F \phi T_{ph} \text{ Volts} \tag{4.41}$$

Where,

$K_C$ is 1 for full pitch coil
$K_d$ is 1 for concentrated winding

### 4.9.4 Voltage Regulation of an Alternator

The voltage regulation is defined as the change in its terminal voltage [15]. The expression for voltage regulation of an alternator is given in equation 4.42.

$$\%\text{Regulation} = \frac{E_{ph} - V_{ph}}{V_{ph}} \times 100\ (\%) \tag{4.42}$$

Where,

$V_{ph}$ is rated terminal voltage
$E_{ph}$ is no load induced EMF

The relationship between the load current and terminal voltage is shown in Figure 4.51.

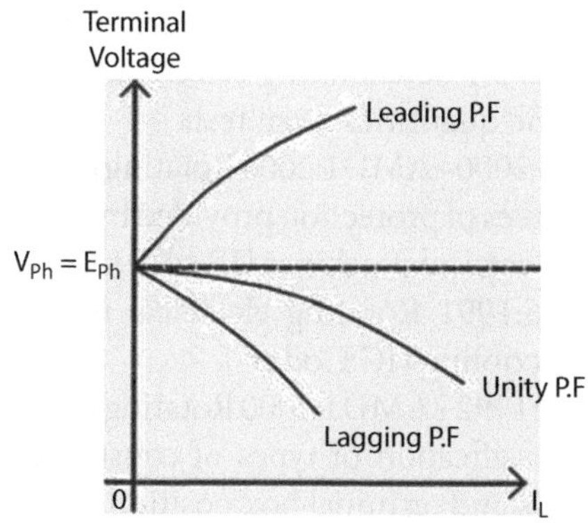


**Figure 4.51** Relationship between the load current and terminal voltage.

## 4.10 Standards for Electric Machines

The standards used for electric machines are as follows:

1. International Electrotechnical Commission (IEC)
2. National Electrical Manufacturers Association
3. Bureau of Indian standards

IEC standards for rotating machines:

1. IEC 60034-1:2017 Rotating electrical machines - Part 1: Rating and performance
2. IEC 60034-2-1:2014 Rotating electrical machines - Part 2-1: Standard methods for determining losses and efficiency from tests (excluding machines for traction vehicles)
3. IEC 60034-2-2:2010 Rotating electrical machines - Part 2-2: Specific methods for determining separate losses of large machines from tests - Supplement to IEC 60034-2-1
4. IEC 60034-2-3:2020 Rotating electrical machines - Part 2-3: Specific test methods for determining losses and efficiency of converter-fed AC motors
5. IEC 60034-3:2007 Rotating electrical machines - Part 3: Specific requirements for synchronous generators driven by steam turbines or combustion gas turbines

6. IEC 60034-4-1:2018 Rotating electrical machines - Part 4-1: Methods for determining electrically excited synchronous machine quantities from tests
7. IEC 60034-5:2000+AMD1:2006 Rotating electrical machines - Part 5: Degrees of protection provided by the integral design of rotating electrical machines (IP code)-Classification
8. IEC 60034-6:1991 Rotating electrical machines - Part 6: Methods of cooling (IC Code)
9. IEC 60034-7:1992+AMD1:2000 Rotating electrical machines - Part 7: Classification of types of construction, mounting arrangements and terminal box position (IM Code)
10. IEC 60034-8:2007+AMD1:2014 Rotating electrical machines - Part 8: Terminal markings and direction of rotation
11. IEC 60034-9:2003+AMD1:2007 Rotating electrical machines-Part 9: Noise limits
12. IEC 60034-11:2004 Rotating electrical machines - Part 11: Thermal protection
13. IEC 60034-12:2016 Rotating electrical machines - Part 12: Starting performance of single-speed three-phase cage induction motors
14. IEC 60034-14:2018 Rotating electrical machines - Part 14: Mechanical vibration of certain machines with shaft heights 56 mm and higher - Measurement, evaluation and limits of vibration severity
15. IEC 60034-15:2009 Rotating electrical machines - Part 15: Impulse voltage withstand levels of form-wound stator coils for rotating a.c. machines
16. IEC 60034-16-1:2011 Rotating electrical machines - Part 16-1: Excitation systems for synchronous machines–Definitions
17. IEC TR 60034-16-2:1991 Rotating electrical machines-Part 16: Excitation systems for synchronous machines - Chapter 2: Models for power system studies
18. IEC TS 60034-16-3:1996 Rotating electrical machines - Part 16: Excitation systems for synchronous machines - Section 3: Dynamic performance
19. IEC 60034-18-1:2010 Rotating electrical machines - Part 18-1: Functional evaluation of insulation systems - General guidelines
20. IEC 60034-18-21:2012 Rotating electrical machines - Part 18-21: Functional evaluation of insulation systems - Test

procedures for wire-wound windings - Thermal evaluation and classification

21. IEC 60034-18-31:2012 Rotating electrical machines - Part 18-31: Functional evaluation of insulation systems - Test procedures for form-wound windings - Thermal evaluation and classification of insulation systems used in rotating machines
22. IEC 60034-18-32:2010 Rotating electrical machines - Part 18-32: Functional evaluation of insulation systems - Test procedures for form-wound windings - Evaluation by electrical endurance
23. IEC TS 60034-18-33:2010 Rotating electrical machines-Part 18-33: Functional evaluation of insulation systems-Test procedures for form-wound windings - Multifactor evaluation by endurance under simultaneous thermal and electrical stresses
24. IEC 60034-18-34:2012 Rotating electrical machines - Part 18-34: Functional evaluation of insulation systems - Test procedures for form-wound windings - Evaluation of thermo mechanical endurance of insulation systems
25. IEC 60034-18-41:2014+AMD1:2019 Rotating electrical machines - Part 18-41: Partial discharge free electrical insulation systems (Type I) used in rotating electrical machines fed from voltage converters - Qualification and quality control tests
26. IEC 60034-18-42:2017 Rotating electrical machines - Part 18-42: Partial discharge resistant electrical insulation systems (Type II) used in rotating electrical machines fed from voltage converters - Qualification tests
27. IEC 60034-19:2014 Rotating electrical machines - Part 19: Specific test methods for d.c. machines on conventional and rectifier-fed supplies
28. IEC TS 60034-20-1:2002 Rotating electrical machines - Part 20-1: Control motors - Stepping motors
29. IEC 60034-22:2009 Rotating electrical machines - Part 22: AC generators for reciprocating internal combustion (RIC) engine driven generating sets
30. IEC 60034-23:2019 Rotating electrical machines - Part 23: Repair, overhaul and reclamation
31. IEC TS 60034-24:2009 Rotating electrical machines - Part 24: Online detection and diagnosis of potential failures at

the active parts of rotating electrical machines and of bearing currents - Application guide

32. IEC TS 60034-25:2014 Rotating electrical machines - Part 25: AC electrical machines used in power drive systems - Application guide
33. IEC 60034-26:2006 Rotating electrical machines - Part 26: Effects of unbalanced voltages on the performance of three-phase cage induction motors
34. IEC 60034-27-1:2017 Rotating electrical machines - Part 27-1: Off-line partial discharge measurements on the winding insulation
35. IEC TS 60034-27-2:2012 Rotating electrical machines - Part 27-2: On-line partial discharge measurements on the stator winding insulation of rotating electrical machines
36. IEC 60034-27-3:2015 Rotating electrical machines - Part 27-3: Dielectric dissipation factor measurement on stator winding insulation of rotating electrical machines
37. IEC 60034-27-4:2018 Rotating electrical machines - Part 27-4: Measurement of insulation resistance and polarization index of winding insulation of rotating electrical machines
38. IEC 60034-28:2012 Rotating electrical machines - Part 28: Test methods for determining quantities of equivalent circuit diagrams for three-phase low-voltage cage induction motors
39. IEC 60034-29:2008 Rotating electrical machines - Part 29: Equivalent loading and superposition techniques - Indirect testing to determine temperature rise
40. IEC 60034-30-1:2014 Rotating electrical machines - Part 30-1: Efficiency classes of line operated AC motors (IE code)
41. IEC TS 60034-30-2:2016 Rotating electrical machines - Part 30-2: Efficiency classes of variable speed AC motors (IE-code)
42. IEC TS 60034-31:2010 Rotating electrical machines - Part 31: Selection of energy-efficient motors including variable speed applications - Application guide
43. IEC TS 60034-32:2016 Rotating electrical machines - Part 32: Measurement of stator end-winding vibration at form-wound windings

**Students and Lecturer conversation**

**1. Lecturer: What is an Electric Motor?**
**Student:** An electric motor is a device which is used to convert electrical energy into rotating mechanical energy. This rotating mechanical energy is used for various applications like fans, drives, water pumping, etc.

**2. Lecturer: What are all the advantages of a three-phase induction motor?**
**Student:** The advantages of three-phase induction motors are:

i) Induction motors have a self-starting property and no need for external starting devices
ii) High power factor
iii) Simple and rugged construction
iv) High reliability
v) Efficiency is high

**3. Lecturer: What are all the types of three-phase induction motor?**
**Student:** The types of three-phase induction motors are:

1. Squirrel cage induction motor
2. Slip ring induction motor

**4. Lecturer: What is slip speed of the induction motor?**
**Student:** The speed difference between the synchronous speed ($N_S$) and actual speed of rotor (N) is called slip speed or slip of an induction motor.

**5. Lecturer: What is the expression for synchronous speed?**
**Student**: The expression for synchronous speed is as follows:

$$N_S = \frac{120 \times F}{P} \text{ in RPM}$$

**6. Lecturer: What are the types of single phase induction motor?**
**Student:** The types of the single phase induction motors are:

1. Split-phase induction motor
2. Capacitor-start induction motor
3. Capacitor-start and capacitor-run induction motor

4. Shaded-pole induction motor

**7. Lecturer: Why are starters used to start the motors?**
**Student:** When a three-phase Induction motor is powered ON at nominal supply voltage, due to zero back EMF at standstill condition, it draws the higher amount of current from the power supply.

**8. Lecturer: What are all the types of starters?**
**Student:** The types of starters are:

i) Direct On Line (DOL) starter
ii) Primary resistor or reactor starter
iii) Auto transformer starter
iv) Star – Delta starter
v) Rotor resistance starter

**9. Lecturer: Up to what capacity can a DOL starter be used?**
**Student:** DOL starters are used for a motor of small capacity, less than 5 kW.

**10. Lecturer: What are all the starting methods used to start the synchronous motor?**
**Student:** The synchronous motor is not a self-starting motor, so in order to start the motor any of the following methods are used:

i). By means of DC motor
ii). By means of AC motor or Pony motor
iii). By means of damper grids in the pole faces

## References

1. D. P. Kothari and I J Nagrath, *Electric Machines*, 4th Edison, Tata McGraw Hill, New Delhi, 2010.
2. S. K. Sahdev, *Electric Machines*, Cambridge University Press, New Delhi, 2018.
3. Saurabh Kumar Mukerji, Ahmad Shahid Khan and Yatendra Pal Singh, *Electromagnetics for Electrical Machines*, CRC Press, Boca Raton, 2015.
4. B. L. Theraja and A. K. Theraja, *Electrical Technology*, volume 2, S. Chand publication, 2006.
5. Jacek F. Gieras, *Electrical Machines*, CRC press, Boca Raton, 2017.
6. Sivaraman, P., and Sharmeela, C. (2020). Introduction to electric distribution system, Baseem Kahn, Hassan Haes Alhelou & Ghassan (Eds.), *Handbook of Research on New Solutions and Technologies in Electrical Distribution Networks*, (pp. 1–31). Hershey, PA: IGI Global.
7. Sivaraman, P., and Sharmeela, C. (2020). Existing issues associated with electric distribution system. In Baseem Kahn, Hassan Haes Alhelou & Ghassan (Eds.), *Handbook of Research on New Solutions and Technologies in Electrical Distribution Networks*, (pp. 1–31). Hershey, PA: IGI Global.
8. Turan Gonen, *Electric* Power *Distribution Engineering*, 3rd Edison, CRC Press, 2014.
9. Juan M Gers, *Distribution system analysis and automation*, IET, 2013.
10. Loi Lei Lai and Tze Fun Chan, *Distributed Generation: Induction and Permanent Magnet Generators*, IEEE Press, 2007.
11. William H. Kersting, *Distribution System Modelling and Analysis*, 4th Edison, CRC press, 2018.
12. IEC 60034-3:2007 Rotating electrical machines - Part 3: Specific requirements for synchronous generators driven by steam turbines or combustion gas turbines.
13. IEC 60034-4-1:2018 Rotating electrical machines - Part 4-1: Methods for determining electrically excited synchronous machine quantities from tests.
14. IEC 60034-12:2016 Rotating electrical machines - Part 12: Starting performance of single-speed three-phase cage induction motors.
15. IEC 60034-16-1:2011 Rotating electrical machines - Part 16-1: Excitation systems for synchronous machines – Definitions.
16. IEC TS 60034-20-1:2002 Rotating electrical machines - Part 20-1: Control motors - Stepping motors.
17. IEC TS 60034-25:2014 Rotating electrical machines - Part 25: AC electrical machines used in power drive systems - Application guide.

# 5

# Measurement and Instrumentation

## Measurement and Instrumentation

A measurement is made by comparing a quantity with a standard unit [1]. Instrumentation is the variety of measuring instruments to monitor and control process variables such as Level, Pressure, Temperature, Humidity, Flow, pH, Force, Speed etc.

## 5.1 Electrical and Electronic Instruments

Electrical instruments depend on the mechanical movement of an indicating device having some inertia and thus have a limited time response [2]. The elements used in such instruments are

1. A detector
2. An intermediate transfer device
3. An induction, a recorder or a storage device.

These are used for the detection of electromagnetic produced signals such as radio, video and microwave space applications and computers.

### 5.1.1 Classification of Instruments

An instrument is a device that is used to do a particular task [3]. It is a device that is used for making measurements like voltage, current, speed, etc. The classification of instruments flowchart is shown in Figure 5.1.

**A. Absolute Instruments**

These instruments give the value of the electrical quantity to measure in terms of the constant and its deflection. These instruments are known as Absolute Instruments.

P. Sivaraman, C. Sharmeela, A. Thaiyal Nayagi, and R. Mahendran. Basic Electrical and Instrumentation Engineering, (243–340) © 2021 Scrivener Publishing LLC

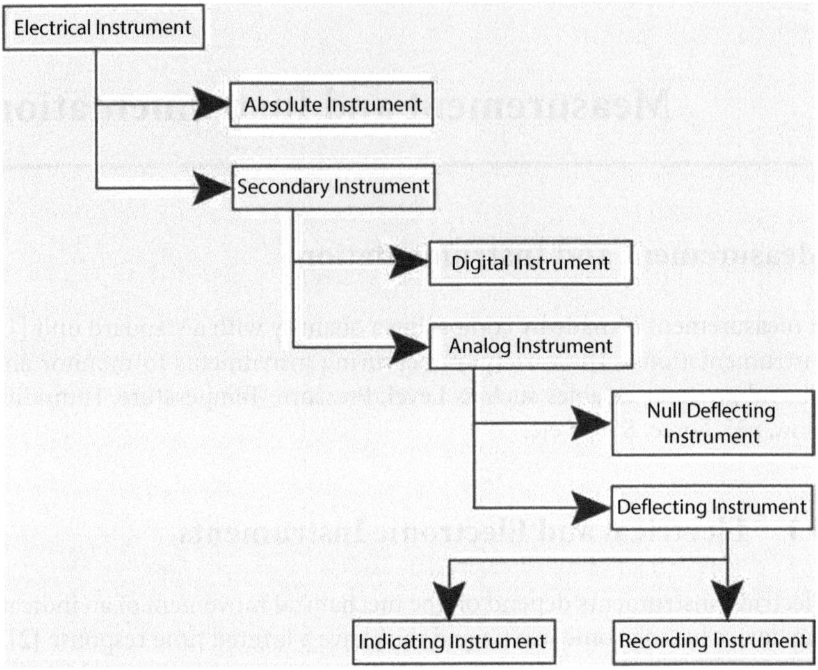


**Figure 5.1** Classification of instruments.

**Examples:**

1. Tangent Galvanometer
2. Rayleigh Current Balance

**i. Tangent Galvanometer**

The tangent galvanometer was first described in an 1837 paper by Claude-Servais-Mathias Pouillet (1790-1868), who later employed this sensitive form of galvanometer to verify Ohm's law.

Tangent Galvanometer is an early measuring instrument for small electric currents [4]. It consists of a coil of insulated copper wire wound on a circular non-magnetic frame. It's working is based on the principle of the tangent low of magnetism [5]. When a current is passed through the circular coil, a magnetic field (B) is produced at the center of the coil in a direction perpendicular to the plane of the coil. The Tangent Galvanometer is shown in Figure 5.2.

**Figure 5.2** Tangent Galvanometer (Courtesy: Tangent Galvonometer, Magnetic Field Measuring Instruments: Eisco Labs).

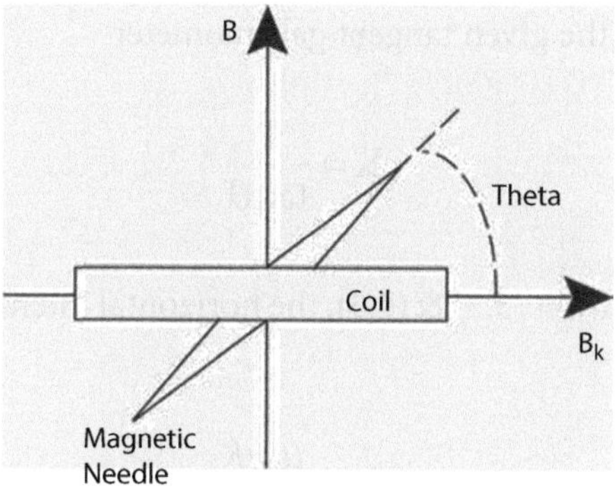


**Figure 5.3** Relationship of B and $B_h$.

When a bar magnet is suspended in two Magnetic fields B and $B_h$, it comes to rest making an angle $\theta$ with the direction of $B_h$ as shown in Figure 5.3.

Here $B = B_h \tan\theta$. This is known as tangent law of magnetism. The Tangent Galvanometer is arranged in such a way that the component of earth's magnetic field $B_h$ is in the direction of the plane of the coil. The magnetic needle is then under the action of two perpendicular fields. If $\theta$ is the deflection of the needle, the according to tangent law

$$B = B_h \tan\theta \tag{5.1}$$

Let I is the current passing through the coil of radius a with n turns, then the field generated by the current carrying circular coil is

$$B = \frac{\mu_o nI}{2a} \tag{5.2}$$

Equating (5.1) and (5.2),

$$B_h \tan\theta = \frac{\mu_o nI}{2a} \tag{5.3}$$

$$\frac{2aB_h}{\mu_o^n} = \frac{I}{\tan\theta} \tag{5.4}$$

The left hand side of equation (5.4) is a constant and is called the reduction factor (K) of the given tangent galvanometer

$$\mathrm{K} = \frac{\mathrm{I}}{\tan\theta} \tag{5.5}$$

Now from equation (5.3) & (5.5), the horizontal intensity of earth's magnetic field $B_h$ is

$$B_h = \frac{\mu_o nK}{2a} \tag{5.6}$$

The applications of tangent galvanometers are:

1. It is used to measure the magnitude of the horizontal component of the geomagnetic field
2. It compares the galvanometer constants
3. It is used to measure current

**ii. Rayleigh's Current Balance**

| Rayleigh's Current Balance inventors are Edward B. Rosa (Edward Bennett), 1861-1921; J. M. Miller; and Noah E. Dorsey at National Institute of Standards and Technology. |
|---|

Absolute instruments give the value of the quantity to be measure in terms of the deflection and instrument constant. The Rayleigh's current balance is an absolute instrument. This instrument is used in standard laboratories and institution as standardizing instruments.

Raleigh's current balance works on the principle of force between two parallel current carrying conductors [6]. Here one coil carrying a current placed with its plane parallel to another coil carrying a current with their axes coincident. Then there is a force exerted between the coils.

The force of attraction between circuits carrying electrical currents and determine the current by measuring the forces of attraction is called the current balance method. The potential energy of two circuits relative to one another is equal to I1, I2 and M.

Where

I1 & I2 is the strength of the currents flowing in the two circuits.
M is Mutual Inductance

The force between the two circuits tending to produce a displacement is measured by the rate of change of potential energy of these determinations

i. A direct determination of the force of restitution of the deformed body as it was used in the electrical measurements
ii. It requires the measurement of a force by weighing and the absolute value of the acceleration of gravity.

Figure 5.4 shows the arrangement of Raleigh's current balance working on the principle of force between two parallel current carrying coils.

From Figure 5.4, there are two fixed coils and a small moving coil. The moving coil is suspended from the arm of a sensitive balance [7]. The three coils are connected in series so current through them remains the same as I.

Then the force exerted on the moving coil is proportional to the square of the current and given by the relation [6].

$$F = I^2 \frac{dM}{dX} N \tag{5.7}$$

Where,

$\frac{dM}{dX}$ is rate of change of mutual inductance between fixed and moving coil in H/m

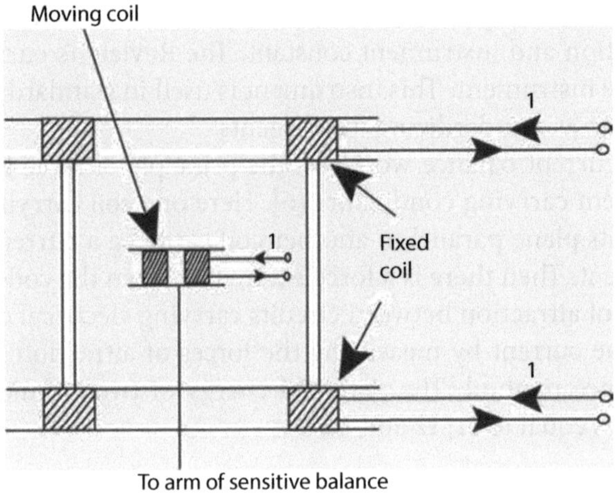


**Figure 5.4** Rayleigh's current balance.

**B. Secondary Instruments**

Secondary instruments give the value of the quantity to be measured directly into the deflection. Secondary Instruments are also called Direct Instruments [8]. These instrument values are required to compare with absolute instruments or standard value of the instruments. It gives the magnitude of the electrical quantity to be measured directly.

i) Every time measurement is made, it takes more time to compute the magnitude of the quantity under measurement.
ii) Secondary instruments are most commonly used.

The Secondary Instruments work under two modes:

1. Analog Mode
2. Digital Mode

**i. Analog Mode**

Analog signals vary in a continuous fashion and take on infinity of values in any given range [12]. Any information may be conveyed by an analog signal. Often such a signal is a measured response to changes in physical phenomena, such as temperature, pressure, sound or position and is achieved using a transducer. An analog signal is one where at each point in time the value of the signal is significant. It is shown in Figure 5.5.

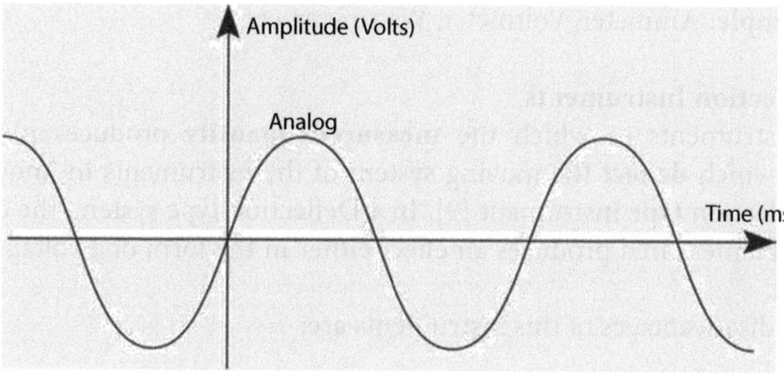


**Figure 5.5** Analog signal.

For example, in sound recording, fluctuations in air pressure strike the diaphragm of a microphone which induces corresponding fluctuations in the current produced by a coil in an electromagnetic microphone or the voltage produced by a condenser microphone.

**ii. Digital Mode**

The digital values are discrete and vary in equal steps. Signals which vary in discrete steps and take up only finite different values in a given range are called Digital signal. This is shown in Figure 5.6.

The secondary instruments are classified into two parts:

i) Deflection Instruments
ii) Null Deflection Instruments

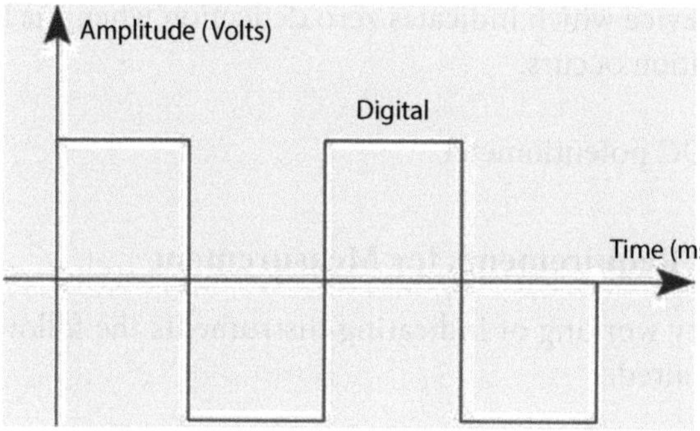


**Figure 5.6** Digital signal.

**Example:** Ammeter, Voltmeter, Wattmeter, etc.

**a) Deflection Instruments**

The instruments in which the **measured quantity** produces physical effects which **deflect** the moving system of the instruments in known as the deflection type instrument [9]. In a Deflection type system, the quantity to be measured produces an effect either in the form of a voltage or a current.

The disadvantages of this instruments are:

1. The accuracy of the instruments is low.
2. The sensitivity of the instruments is less than the null type instruments.

**b) Null Deflection Instruments**

An instrument in which zero or null indication determines the magnitude of measured quantity is called a null type instrument [9]. It uses a null detector which indicates the null condition when the measured quantity and the opposite quantity are the same. It has high accuracy and also it is very sensitive.

For the operation of the null type instrument, the following conditions are required:

1. The effect produced by the measuring quantity.
2. The opposing effects whose values are accurately known. It is necessary for measuring for determining the numerical value of the measured quantity accurately.
3. A detector which detects the null conditions, i.e., a detector is a device which indicates zero deflection when the balance condition occurs.

**Example:** DC potentiometer

### 5.1.2 Basic Requirements for Measurement

For satisfactory working of indicating instruments the following types of torque are required.

i. Deflection torque (operating)
ii. Controlling (restoring) torque
iii. Damping torque

**A. Deflection Torque (Operating)**

It is the torque which deflects the pointer on a calibrated scale according to the electrical quantity passing through the instrument [10]. The deflecting torque causes moving system and hence pointer attached to it moves from zero position to indicate electrical quantity being measured on graduated scale.

$$\text{Deflecting Torque} \propto \text{Measurable quantity}$$

The deflecting torque is produced by utilizing one or more of the following effects of current or voltage:

1. Magnetic Effect --> Moving Iron Instruments
2. Electro-dynamic Effect-->(i) Moving Coil Instruments
   (ii) Dynamometer Type
3. Electro-magnetic Induction Effect--> Induction Type Instruments
4. Thermal Effect -->Hot-Wire Instruments
5. Chemical Instrument -->Electrolytic Instruments
6. Electro-static Effect-->Electrostatic Voltmeters

The information about the electrical measuring instruments in which deflecting torque is produced by utilizing the first three effects are listed in Table 5.1.

It is necessary to make the moving system of the instrument (pointer) move from its zero position.

$$T_{dc} = NBild = Gi \tag{5.8}$$

$$T_{dc} \ \alpha \ \text{measured}$$

Where,

N is Number of turns of coil
B is Flux Density in air gap at the coil position
L is Length of the vertical sides of coil
D is length of horizontal side of coil
I is Current through the coil
G is NBld displacement constant of the galvanometer

**Table 5.1** Information about the electrical measuring instruments.

| S. No. | Type | Effect | Suitable for | Instrument |
|---|---|---|---|---|
| 1 | Moving-iron | Magnetic effect | DC and AC | Ammeters |
| | | | | Voltmeters |
| 2 | Moving coil | Electro-dynamic effect | DC only | Ammeters |
| | | | | Voltmeters |
| 3 | Dynamometer type | Electro-dynamic effect | DC and AC | Ammeters |
| | | | | Voltmeters |
| | | | | Watt-meters |
| 4 | Induction type | Electromagnetic Induction effect | AC only | Ammeters |
| | | | | Voltmeters |
| | | | | Watt-meters |
| | | | | Energy meters |

**B. Controlling (Restoring) torque**

It is the torque which controls movement of the pointer on a particular scale according to the quantity of electricity passing through it [10]. If deflecting torque acts alone, the pointer would continue to move indefinitely and would swing over to the maximum deflected position irrespective of the magnitude of the current to be measured.

The controlling torque is opposite to deflecting torque. When deflecting torque is equal to controlling torque, the pointer comes to final steady-state position.

- At equilibrium, $T_C = T_D$
- Controlling torque also used to bring the pointer in zero initial position, if there is no deflecting torque
- Controlling torque is provided by spring control and gravity control

Some controlling force is employed either by a spring or gravity to limit the movement.

(a) Spring is generally used to control torque (Phosphor Bronze is the most suitable material). The controlling torque is shown in Figure 5.7.

$$T_C = K\theta \tag{5.9}$$

Where,
$\theta$ is deflection of the pointer
K is controlling torque constant

(b) Gravity method is also sometimes used for controlling torque.

$$T_C \, \alpha \, \sin\theta \tag{5.10}$$

The controlling torque in indicating instruments is provided by either of the following two methods:

- Spring control
- Gravity control

**Spring Control**

This is the most common method of providing controlling torque in electrical instruments [10]. A spiral hairspring made of some non-magnetic

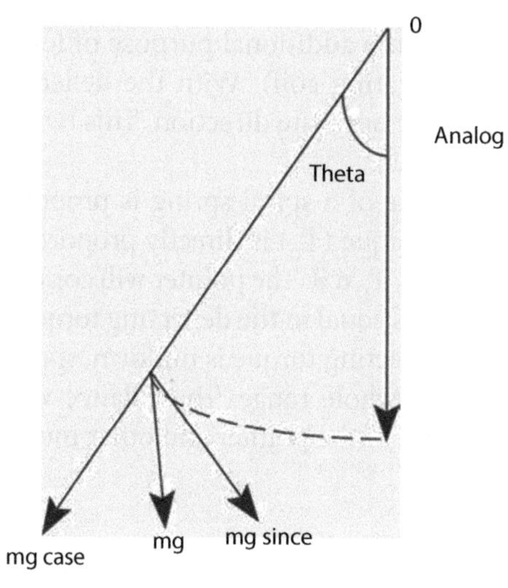


**Figure 5.7** Controlling torque due to gravity.

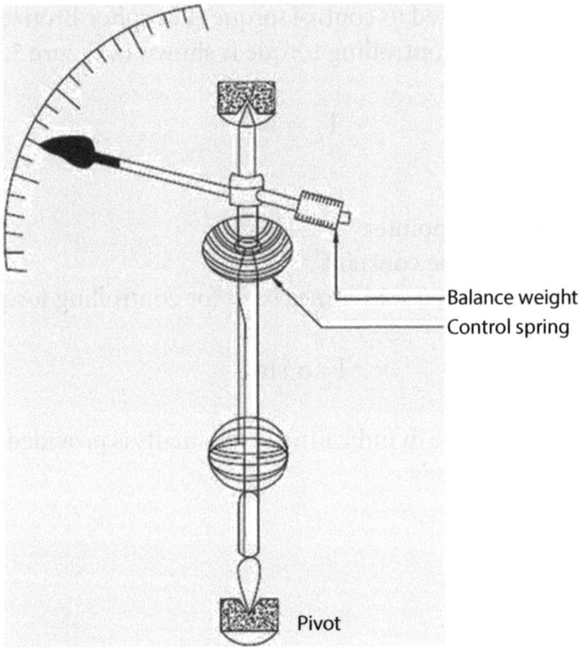


**Figure 5.8** Spring control.

material like Phosphor bronze is attached to the moving system of the instrument as shown in Figure 5.8.

The spring also serves the additional purpose of leading current to the moving system (i.e., operating coil). With the deflection of the pointer, the spring is twisted in the opposite direction. This twist in the spring provides the controlling torque.

Since the torsion torque of a spiral spring is proportional to the angle of twist, the controlling torque ($T_C$) is directly proportional to the angle of deflection of point ($\theta$)., i.e., $T_C \alpha \theta$. The pointer will come to rest at a position where controlling torque is equal to the deflecting torque, i.e., $T_D = T_C$. In an instrument where the deflecting torque is uniform, spring control provides a uniform scale over the whole range. The balance weight is attached to counterbalance the weight of the pointer and other moving parts.

**Gravity Control Method**

In this method, a small weight is attached to the moving system, which provides the necessary controlling torque. In the zero position of the pointer,

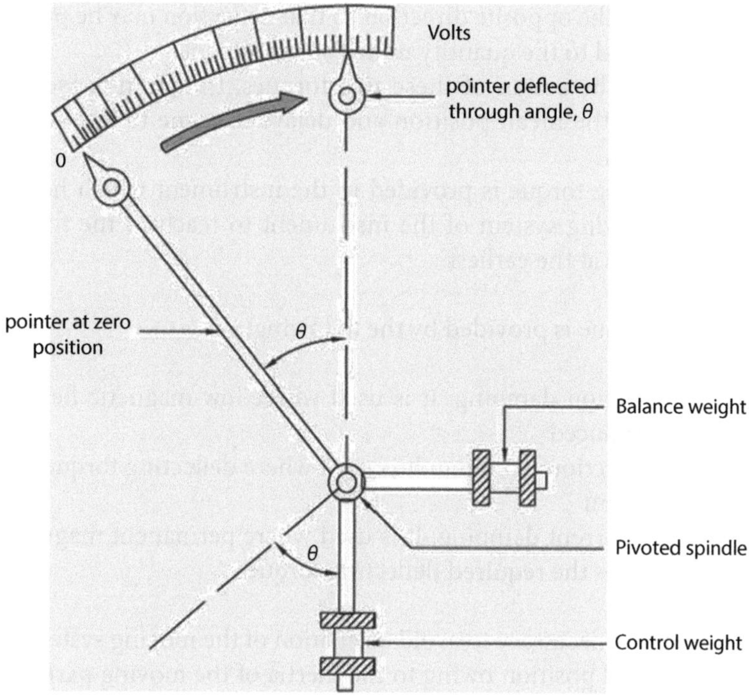


**Figure 5.9** Gravity control.

the control weight hangs vertically downward and therefore it provides the controlling torque. The Gravity Control is shown in Figure 5.9.

However, under the action of deflecting torque, the pointer moves from zero position and the control weight moves in opposite direction. Due to gravity, the control weight would tend to come in the original position (i.e., vertical) and thus provides a controlling torque. The pointer comes to rest at a position where controlling torque is equal to the deflecting torque. In this method, controlling torque ($T_C$) is proportional to the sine of angle of deflection ($\theta$) i.e.$T_C$ α Sin$\theta$. Because in this method, controlling torque ($T_C$) is not directly proportional to the angle of deflection ($\theta$) but it is proportional to Sin $\theta$ therefore, gravity control instruments have non-uniform scales.

**C. Damping Torque**

This torque is used to damp out oscillation at the final steady-state position [10]. The time response of the instrument depends on damping torque.

- Under the action of deflecting torque, moving system of the instrument defects, while the controlling torque is applied

on it in the opposite direction so that deflection may be proportional to the quantity under measurement
- Under the action of these two torques, the pointer oscillates at the mean position and delays to come to the final position
- Damping torque is provided in the instrument which help the moving system of the instrument to reach to the final position at the earliest

Damping torque is provided by the following:

1. Air friction damping: It is used where low magnetic fields are produced
2. Fluid friction damping: It is used where deflecting torque is minimum
3. Eddy current damping: It is used where permanent magnet produces the required deflecting torque.

This torque is necessary to avoid oscillation of the moving system about its final deflected position owing to the inertia of the moving parts and to bring the moving system to rest in its deflected position quickly, as shown in Figure 5.10.

$$T_{dm} = D\frac{d\theta}{dt} \tag{5.11}$$

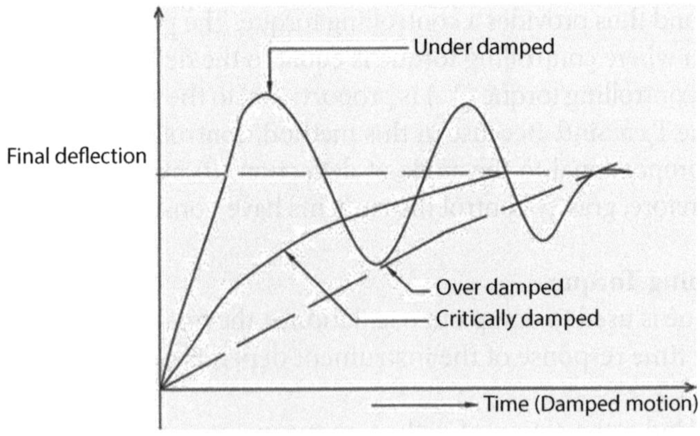


**Figure 5.10** Time vs. final deflection.

Where,

D is Damping Constant

The restarting torque due to inertia of the moving system is

$$T_i = J\left(\frac{d^2\theta}{dt}\right) \tag{5.12}$$

Where,

J is inertia constant

### Types of Damping Torque

The damping torque is classified into three types:

- Air Friction damping
- Fluid Friction damping
- Eddy Current damping

### i. Air Friction damping

A light aluminum piston is attached to the spindle that carries the pointer and moves with very little clearance in a rectangular air chamber closed at one end [11]. The cushioning action of the air on the piston damps out any tendency of the pointer to oscillate about the final deflected position as shown in Figure 5.11.

In this method, one or two light aluminum vanes are attached to the same spindle that carries the pointer. As the pointer moves, the vanes swing and compress the air. The pressure of compressed air on the vanes provides the necessary damping force to reduce the tendency of the pointer to oscillate.

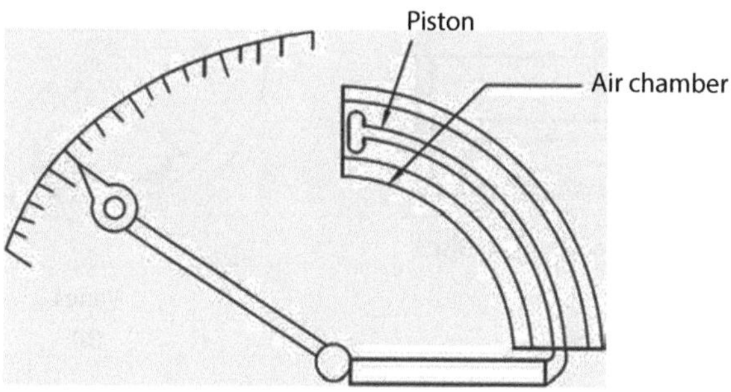


**Figure 5.11** Air friction damping.

### ii. Fluid Friction Damping

In this method, discs or vanes attached to the spindle of the moving system are kept immersed in a pot containing oil of high viscosity. As the pointer moves, the friction between the oil and vanes opposes the motion of the pointer and thus necessary damping is provided as shown in Figure 5.12 a & b.

The fluid friction damping is not suitable for portable instruments because of oil contained in the instrument. In general, fluid friction damping is not employed in an indicating instrument, although one can find its use in a kelvin electrostatic voltmeter.

### iii. Eddy Current Damping

When a conductive material is subjected to a time-varying magnetic flux, eddy currents are generated in the conductor. Two methods of eddy current damping are generally used [11]. The first method is shown in Figure 5.13; a thin aluminum or copper disc attached to the moving system is allowed to pass between the poles of the permanent magnet.

As the pointer moves, the disc cuts across the magnetic field and eddy currents are induced in the disc. These eddy currents react with the field and eddy magnet to produce a force which opposes the motion according to Lenz's Law. In this way, eddy current damping torque reduces the oscillations of the pointer. In the second method, the coil which produces the deflecting torque is wound on aluminum former. As the coil moves in the field of the instrument, eddy currents are induced in the aluminum former to provide the necessary damping torque.

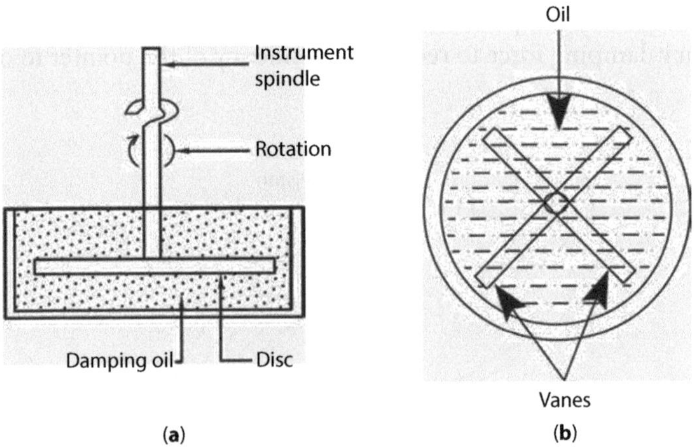


**Figure 5.12** (a) Fluid friction damping. (b) Fluid friction damping.

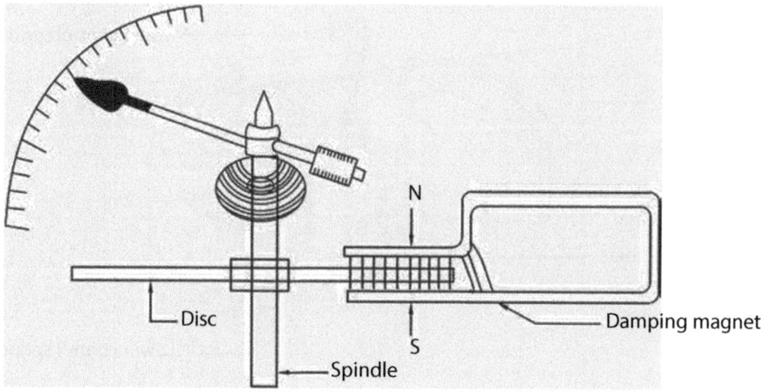


**Figure 5.13** Eddy friction damping.

### 5.1.3 Types of Indicating Instruments

Indicating instruments can be divided into different types according to their working principle.

**A. D'Arsonval Movement**

A commonly used sensing mechanism in DC ammeters, voltmeters and Ohm meters as a current sensing device is called D'Arsonval meter movement [2].

> The most common type of meter movement is the d'Arsonval movement, which was invented by the French physicist Jacques d'Arsonval (1851–1940). This meter movement is based on the interaction of the fields of a permanent magnet and an electromagnet.Imus

When an electric current is passed through a coil placed in a magnetic field, it experiences a force. This force causes a torque in the coil that is fixed to a spindle. The spindle can rotate in fixed bearings. The rotation of the spindle is proportional to the electric current passed through the coil. This torque is balanced after the movement against the restoring torques of springs. The torque that is produced tends to rotate the spindle is termed as D'Arsonval Movement. The Figure 5.14 shows the D'Arsonval movement.

The core is suspended between the shaped pole pieces of a permanent magnet, carried a small mirror which serves as an optical pointer and indicates the coil position by reflecting light beam onto a fixed scale [2].

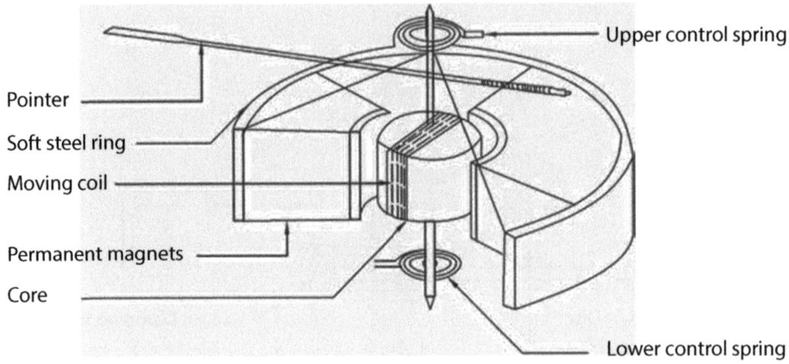


**Figure 5.14** D'Arsonval movement.

The current is measured directed through the coils of the electromagnet so that the magnetic field produced by the current opposes the field of the permanent magnet and causes rotation of the core. The core is restrained by springs so that the needle will deflect or move in proportion to the current intensity. The more current applied to the core, the stronger the opposing field, and the larger the deflection, up to the limit of the current capacity of the coil. When the current is interrupted, the opposing field collapses and the needle is returned to zero by the restraining springs. The limit of the current that can be applied to this type movement is usually less than one mill ampere. A common variation of the D'Arsonval movement is the Weston movement, which uses essentially the same principle built to a more rugged construction by employing jeweled supports for the core and employing a heavier winding in the electromagnet. D'Arsonval movement is a DC device and can only measure DC current or AC current rectified to DC.

The basic permanent magnet moving coil (PMMC) mechanism is often called the D'Arsonval movement. PMMC instruments are accurate and suitable for dc measurements only. When the current (I) passes through a coil, a deflecting torque is produced on the coil.

$$T_{dc} = K_1 I \tag{5.13}$$

The deflecting torque increases the restoring torque in the spring attached to the pointer.

$$T_c = K_2 \theta \tag{5.14}$$

These instruments require very low power consumption and current for full-scale deflection.

The sensitivity of a galvanometer can be specified in terms of current sensitivity and voltage sensitivity.

$$\text{Current Sensitivity } S_I = \frac{d}{I} mm/\mu A \tag{5.15}$$

Where,

d is deflection of the galvanometer in mm
I is Galvanometer current in $\mu A$

$$\text{Voltage Sensitivity } S_v = \frac{d}{V} mm/mV \tag{5.16}$$

V is voltage applied to the galvanometer in mV

**B. DC Ammeter**

The instrument which is used to measure the Direct Current is called DC ammeter as shown in Figure 5.15.

If the electric charge flows only in one direction, then the resultant current is called Direct Current (DC) [13]. It flows in the same direction at all points in time, although the instantaneous magnitude of the *current* varies. Here the resistor ($R_{sh}$) connected parallel with the PMMC. The $R_{sh}$ also called shunt resistance. The value of this resistance measure the DC current of large value. The voltage across shunt resistor ($R_{sh}$) and the voltage

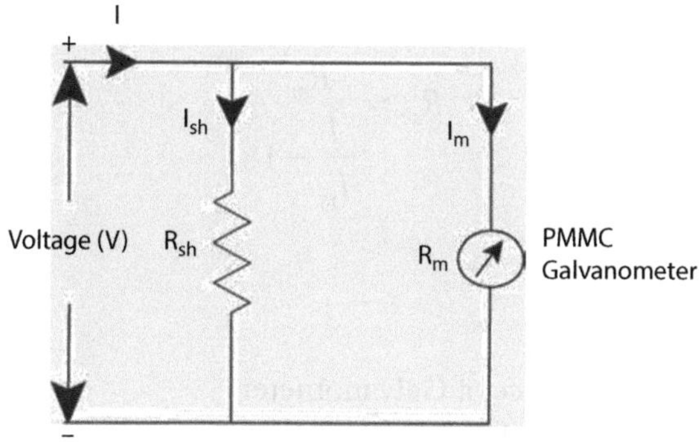


**Figure 5.15** DC ammeter.

across galvanometer resistance ($R_m$) is same, both resistance connected in parallel.

Mathematically, it can be written as

$$I_{sh}R_{sh} = I_m R_m$$

$$R_{sh} = \frac{I_m R_m}{I_{sh}} \quad (5.17)$$

The KCL equation at node 1 is

$$-I + I_{sh} + I_m = 0$$

$$I_{sh} = I - I_m$$

Substitute the value of $I_{sh}$ in Equation 5.17

$$R_{sh} = \frac{I_m R_m}{I - I_m} \quad (5.18)$$

Take, $I_m$ as common in the denominator term, which is present in the right-hand side of equation (5.18)

$$R_{sh} = \frac{I_m R_m}{I_m\left(\frac{I}{I_m} - 1\right)}$$

$$R_{sh} = \frac{R_m}{\frac{I}{I_m} - 1} \quad (5.19)$$

Where,

$R_{sh}$ is Shunt Resistance
$R_m$ is Internal Resistance of Galvanometer
I is Total Direct Current
$I_m$ is Full Scale Deflection Current

The ratio of total Direct Current that is to be measured, I and the full-scale deflection current of the galvanometer, $I_m$ is known as multiplying factor m.

Mathematically, it can be represented as

$$m = \frac{I}{I_m} \tag{5.20}$$

$$R_{sh} = \frac{R_m}{m-1} \tag{5.21}$$

The value of shunt resistance is obtained from equation 5.21.

The coil winding of the PMMC movement is small and light and it can carry only small currents.

**C. Multi-Range DC Ammeter**

The current range of the DC ammeter can be further extended by a number of shunts selected by a range switch; such a meter is called multi-range ammeter and is shown in Figure 5.16.

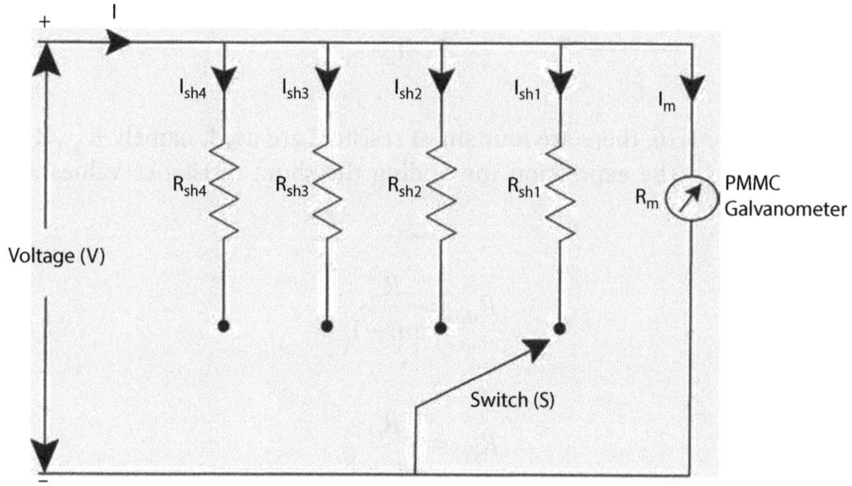


**Figure 5.16** Multi-range DC ammeter.

The DC ammeter for measuring the Direct Currents of multiple ranges, multiple parallel resistors are used instead of a single resistor and this entire combination of resistors is in parallel to the PMMC galvanometer.

This multi-range DC ammeter in series with the branch of an electric circuit, where the Direct Current of required range is to be measured. The desired range of currents is chosen by connecting the switch (S) to the respective shunt resistor.

Let, $m_1$, $m_2$, $m_3$ and $m_4$ are the multiplying factors of DC ammeter when the considered total Direct Currents to be measured as $I_1$, $I_2$, $I_3$ and $I_4$ respectively.

Following are the formulae corresponding to each multiplying factor.

$$m_1 = \frac{I_1}{I_m}$$

$$m_2 = \frac{I_2}{I_m}$$

$$m_3 = \frac{I_3}{I_m}$$

$$m_4 = \frac{I_4}{I_m}$$

In Figure 5.16, there are four shunt resistors are used, namely $R_{sh1}$, $R_{sh2}$, $R_{sh3}$ and $R_{sh4}$. The expression for finding the shunt resistance values are given below

$$R_{sh1} = \frac{R_m}{m_1 - 1}$$

$$R_{sh2} = \frac{R_m}{m_2 - 1}$$

$$R_{sh3} = \frac{R_m}{m_3 - 1}$$

$$R_{sh4} = \frac{R_m}{m_4 - 1}$$

The above formula, find the value of each resistor.

**D. DC Voltmeter**

The basic D'Arsonval movement can be converted into a dc voltmeter with the addition of a series resistor or multiplier as shown in Figure 5.17.

The multiplier limits the current through the movement cannot exceed the value of the full-scale deflection current ($I_{fsd}$) [13].

A DC voltmeter connected across the source of EMF or a circuit component. The terminals are generally marked positive and negative since polarity must be observed.

$$V = I_m(R_s + R_m) \text{ or } \mathrm{R_s} = \frac{\mathrm{V}}{\mathrm{I_m} - \mathrm{R_m}}$$

$$m = \frac{V}{V_m} = I_m(R_m + R_s) \Big/ I_m R_m$$

$$m = 1 + \frac{R_s}{R_m}$$

$$\mathrm{R_s} = (\mathrm{m} - 1)\mathrm{R_m}$$

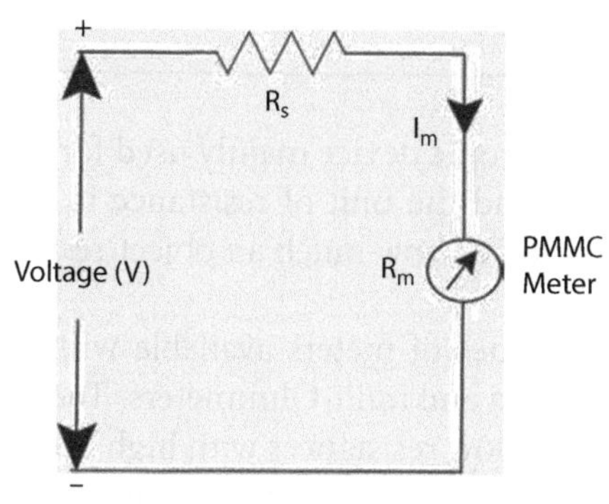


**Figure 5.17** DC voltmeter.

Where,

$I_m$ is deflection current of meter
$R_s$ is multiplier resistance
$V$ is Full-scale voltage of the instruments
$V_m$ is voltage across movement

**E. Multi-Range Voltmeter**

A multi-range voltmeter can be obtained by the addition of a number of multipliers together with the range switch [13]. The Multi-Range Voltmeter is shown in Figure 5.18 (i) and (ii).

Voltmeter sensitivity

$$\left(\Omega/V_{rating}\right), S = \frac{1}{I_{Fsd}} \frac{\Omega}{V}$$

The total resistance of the voltmeter is

$$R_{total} = S \times V$$

$$R_s = (S \times V) - R_m$$

**F. Ohm meter**

> An instrument used for measuring directly the resistance of a conductor or of any part of a circuit through which a strong current is passing. It is the invention of Professor William Edward Ayrton, of London.

An ohmmeter is an electronic device mainly used for calculating electrical resistance of a circuit, and the unit of resistance is Ohm [13]. Electrical resistance is a calculation of how much an object resists allowing the flow of current through it.

There are different types of meters available with different sensitivity levels such as micro, mega and milli-Ohmmeters. The micro-Ohmmeter is used for calculating very low resistances with high precision at specific test currents. This Ohmmeter is used in bonding contact applications.

The basic Ohmmeter circuit diagram is shown Figure 5.19a.

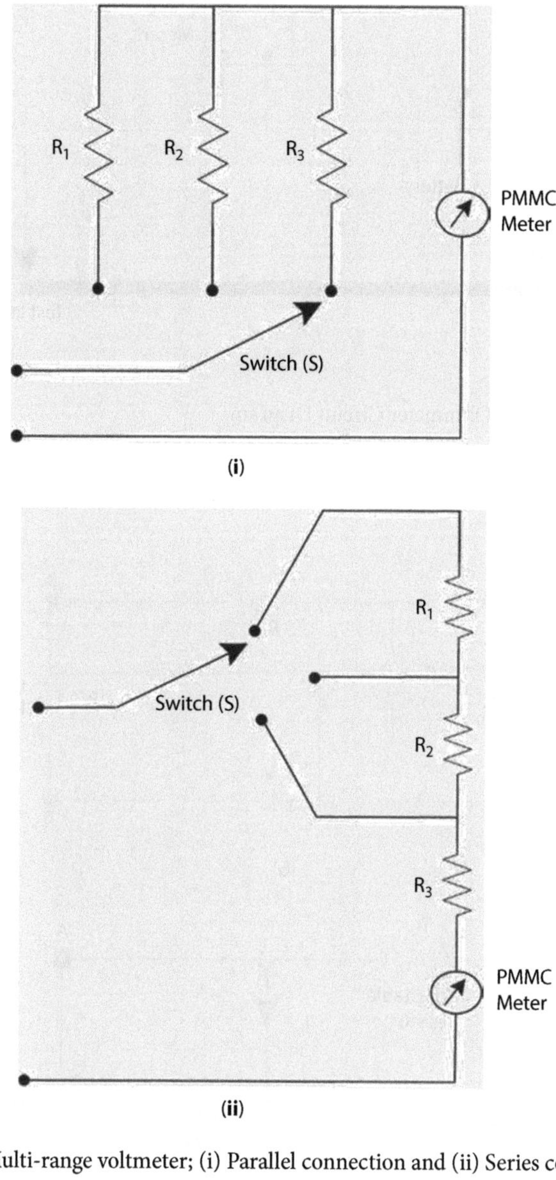


**Figure 5.18** Multi-range voltmeter; (i) Parallel connection and (ii) Series connection.

The Ohmmeter comprises a needle and two test leads. The needle deflection can be controlled with the battery current. Initially, the two test leads of the meter can be shorted together to calculate the resistance of an electrical circuit. Once the two leads of the meter are shorted, then the meter can be changed for appropriate action in a fixed range. The needle

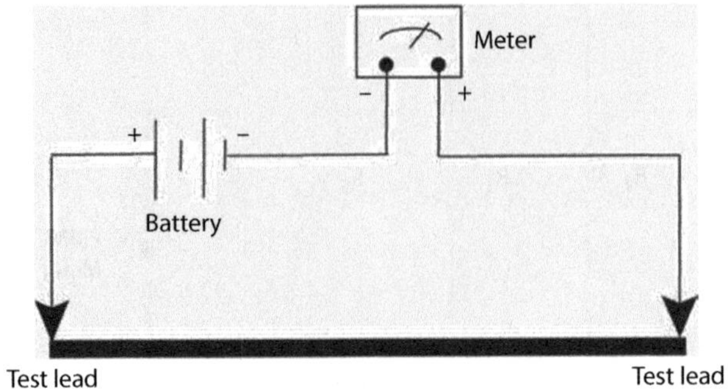


(a)

**Figure 5.19** (a) Basic Ohmmeter Circuit Diagram.

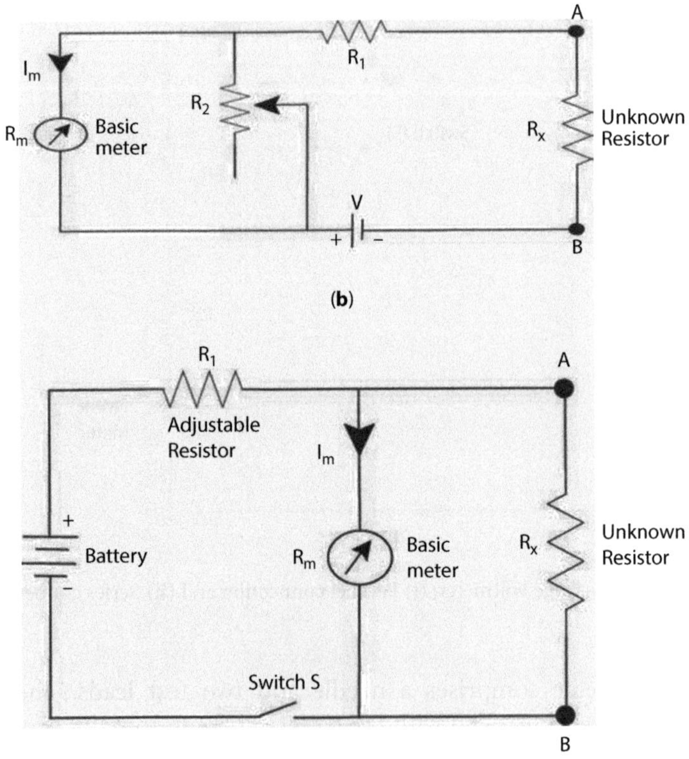


(c)

**Figure 5.19** (b) Series type Ohmmeter. (c) Shunt type Ohmmeter.

comes back to the highest point on the meter scale and the current in the meter will be high.

The Ohm meter is divided into two types:

i. Series type Ohmmeter
ii. Shunt type Ohmmeter

**i. Series type Ohmmeter**

It consists of D'Arsonval movement connected in series with the resistance and a battery to a pair of terminals to which the unknown resistance is connected. The current depends on the magnitude of the unknown resistor $R_x$. The Series Type Ohmmeter is shown in Figure 5.19b.

When $R_X$ is zero, maximum current flows in the circuit and shunt resistance. $R_X$ is adjusted until the movement indicates full-scale current.

A convenient quantity to use in the design of a series type Ohmmeter is the value of $R_x = R_h$ which causes half-scale deflection of the motor.

$$R_h = R_1 + R_2||R_m$$

Then, total resistance for the battery is 2 $R_n$.

$$R_{total} = S \times V$$

Where,

$I_t$ is current through Rx for producing full-scale meter deflection
$I_h$ is current for producing half-scale meter deflection
$I_{fsd}$ is current through movement causing full-scale meter deflection
$I_2$ is current flow through resistor $R_2$
Also, voltage across shunt ($R_2$) is equal to voltage across movement.

$$(R_m) \text{ i.e., } E_{sh} = E_m$$

$$I_2R_2 = I_{Fsd}R_m$$

$$R_2 = I_{Fsd}R_mR_h/(E - I_{Fsd}R_h)$$

$$R_1 = R_h - \frac{I_{Fsd} R_m R_h}{E}$$

**ii. Shunt type Ohm meter**

When $R_X$ is zero, then meter current is zero. If $R_X$ is infinity, a current finds a path through the meter and by appropriate selection of the value of $R_1$, the pointer can be made to read full scale. The shunt type Ohm meter is shown in Figure 5.19c.

When $R_X$ is zero, the full-scale meter current will be

$$I_{Fsd} = \frac{V}{R_1 + R_m}$$

For any value of $R_x$,

$$I_m = \frac{E \times R_x}{(R_1 R_m + R_x (R_1 + R_m))}$$

$$S = \frac{I_m}{I_{Fsd}}$$

$$S = \frac{R_x}{\left(R_x + R_1 R_m / (R_1 + R_m)\right)}$$

$$S = R_x / (R_x + R_P)$$

$$R_p = R_1 || R_m$$

### 5.1.4 AC Indicating Instruments

PMMC meters not be used for measuring AC signals because the meter will show the average value of AC. i.e., Zero Deflection.

### A. Electro Dynamometer

The Electro dynamometer is an instrument which is used for measuring the electric voltage, current and power.

> An electrodynamometer is an instrument used for measuring the electric power. The basic principle was laid out in an 1848 paper by Wilhelm Weber (1804-1891). When the same current passes through two concentric coils placed at right angles to each other, the resulting torque depends on the square of the current.

It contains two types of coils, a fixed coil and a moving coil [2]. The field is produced by a fixed coil and divided in more uniform field as shown in Figure 5.20.

$$\text{Deflecting torque Td} = \text{BANI}$$

Where,

B is Flux Density
A is Area of cross section of coil
I is Current
N is Number of turns of coil

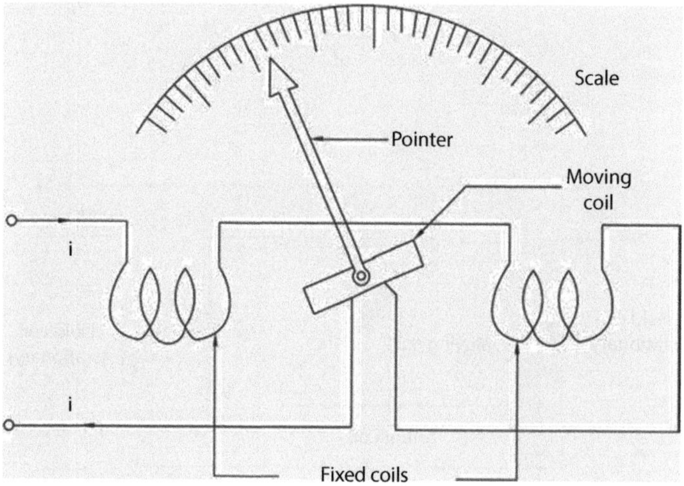


**Figure 5.20** Electrodynamometer.

**i. Electrodynamometer in current and voltage measurement**

In an electrodynamometer, current under measurement produces the magnetic field flux in which the movable coil rotates. The electrodynamometer ammeter circuit diagram is shown in Figure 5.21.

A fixed coil (FC) splits into equal halves and provides the magnetic field in which the moveable coil (MC) rotates.

$$\text{Meter deflection } \alpha\sqrt{(\text{Average} \times \text{i}^2)}$$

In ammeter, fixed and moveable coils are connected in series and therefore carry the same current hence,

$$I_1 = I_2 = I_1, \phi = 0$$

$$\theta = \left(\frac{I^2}{K}\right) \times \frac{d_m}{d_\theta}$$

In voltmeter, the fixed and moveable coils are connected in series with a high non-inductive resistance.

Hence,

$$I_1 = I_2 = \frac{V}{Z}, \ \phi = 0$$

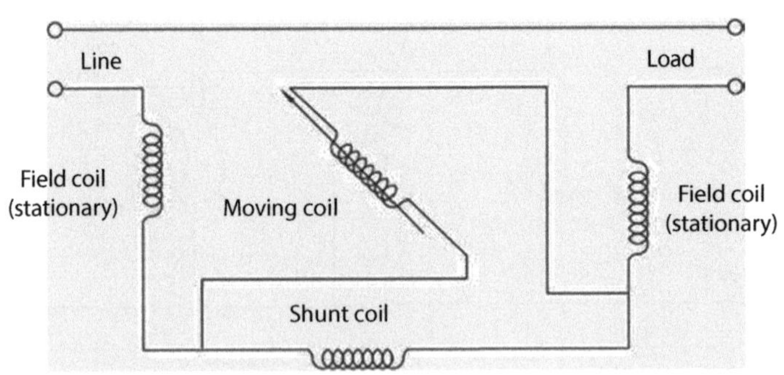


**Figure 5.21** Electrodynamometer ammeter circuit.

$$\theta = \left(\frac{V^2}{Z^2 \times K}\right) \times \frac{d_m}{d_\theta}$$

**ii. Electrodynamometer in power measurement**

The Electrodynamometer is an instrument used for measuring the electric power. The Electrodynamometer in power measurement is shown in Figure 5.22.

It contains two types of coils, a fixed coil and moving coil. The field is produced by a fixed coil and divided in more uniform field. It is used to indicate both AC and DC power for any waveform of voltage and current and is not restricted to a sinusoidal waveform.

$$i_p = \frac{e}{R_p}$$

Where,

e is instantaneous voltage across the power line

$R_p$ is total resistance of the moveable coil

The deflection of moveable coil is proportional the product of the two currents $i_c$ and $i_p$.

Average deflection over one period is

$$Q_{av} = (K/T) \int i_c i_p d_t$$

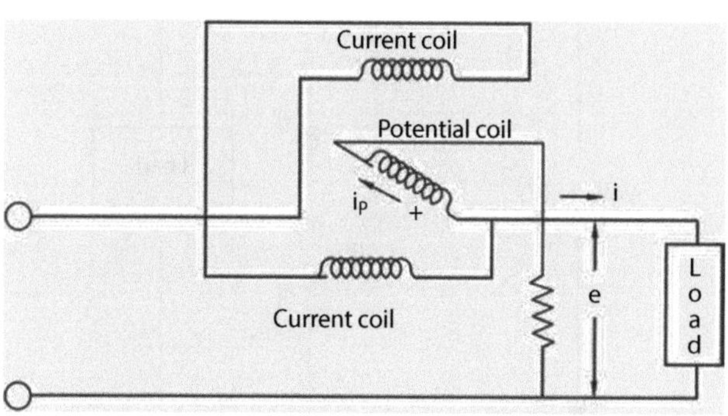


**Figure 5.22** Electrodynamometer in power measurement.

Average power in circuit

$$P_{av} = \frac{1}{T} \int e_i d_t$$

$$P_{av}^{\alpha} Q_{av}$$

**iii. Electrodynamometer Wattmeter**

Electrodynamometer Wattmeter is depends on the reaction between the magnetic field of moving coils and fixed coils. It is used for measurement of power in both AC and DC circuits. The Electrodynamometer Wattmeter is shown in Figure 5.23.

In electrodynamometer Wattmeter, a current carrying conductor is placed in the magnetic field and it experiences the mechanical force. This mechanical force deflects the pointer which is mounted on the Calibrated Scale. An electrodynamometer Wattmeter has a current coil and pressure coil as shown in Figure 5.23.

The instantaneous power is given by

$$T_i = i_1 i_2 \frac{d_m}{d_\theta}$$

Where,

$i_1$ and $i_2$ are the instantaneous currents

m is mutual inductance of the coils

The average deflecting torque is given by

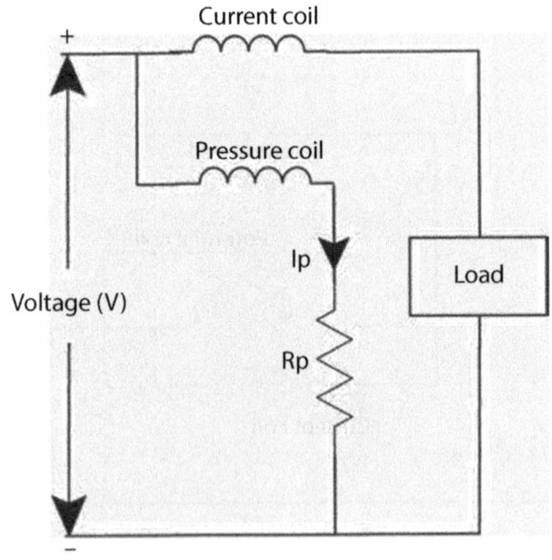


**Figure 5.23** Electrodynamometer Wattmeter.

$$T_i = \left(\frac{Vi_c}{R_p}\right)\cos\phi\frac{d_m}{d_\theta}$$

Where,

$\phi$ is phase angle

$i_c$ is the current coil current

$R_p$ is the series resistance of pressure coil circuits.

$$\theta = \left({VI_c}/{KR_p}\right)\cos\phi\frac{d_m}{d_\theta}$$

If the pressure coil has some inductance, then

$$W = \left({VI_c}/{KR_p}\right)\cos(\phi-\alpha)\cos(\phi-\alpha)\cos\alpha\frac{d_m}{d_\theta}$$

$$W_{true} = \left({VI_c}/{KR_p}\right)\cos\phi\frac{d_m}{d_\theta}$$

$${W}/{W_{true}} = \cos\alpha[\cos\alpha + (\sin\alpha)(\tan\phi)]$$

### 5.1.5 Electrical Instruments

An electrical instruments are the instruments which are used to measure electrical parameters like voltage, current, power, energy, power factor, frequency, etc.

#### A. Multi-meter

> Multi-meters were invented in the early 1920s. The invention of the first multi-meter is attributed to British Post Office engineer Donald MacAdie. He Invented an Instrument which could measure Amps, Volts and Ohms. So the multifunctional meter was then named Avometer.

A Multi-meter or a Multi-tester, also known as a VOM (Volt-Ohm-Milli ammeter) is an electronic measuring instrument that combines several measurement functions in one device. An Electrical Multi-Meter (EMM) is capable of measuring DC and AC voltage, current and resistance.

The characteristics of multi-meters are as follows,

- The input impedance is very high, varying from 1 to 10 MΩ
- Electrical multi-meter can respond to high frequency of 3 MHZ
- The voltage placed across the components during the resistance measurement can be lower than the voltage from a non-electronic multi-meter

**i. Scheme of DC voltage measurement**

The Block diagram of DC Voltage Measurement is shown in Figure 5.24.

The input DC voltage is apply to the attenuator. The attenuator is simply a voltage divider that divides high input voltages to measurable levels. The output of the attenuator is given to DC amplifier. It amplifies the AC signal by low frequency. The output of DC amplifier given to the PMMC meter, it displays the output.

**ii. Scheme of AC voltage measurement**

The AC voltage sources vary from fluctuation in AC voltage over power distribution systems. The voltage measured in AC power systems should be within -10% and +5%. The Block diagram of AC Voltage Measurement is shown in Figure 5.25.

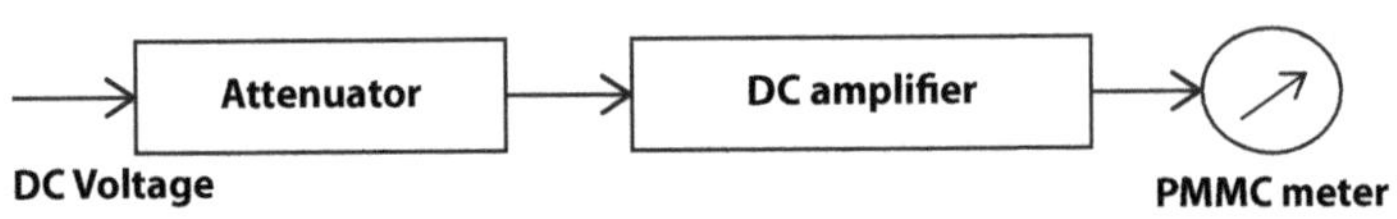


**Figure 5.24** DC voltage measurement.

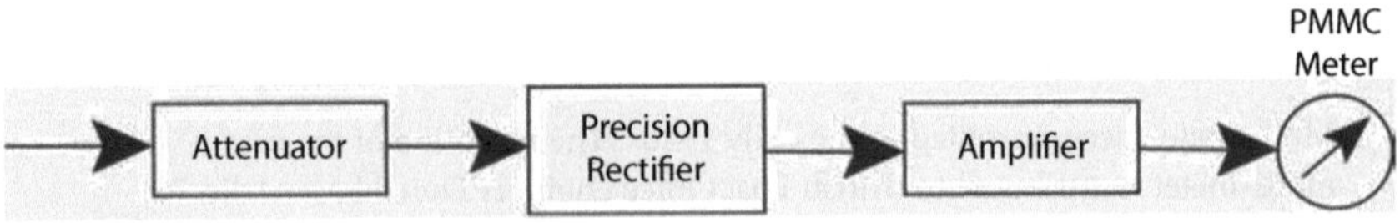


**Figure 5.25** AC voltage measurement.

The output of the attenuator is given to rectifier which converts the AC voltage into pulsating DC voltage. Then the final output of the DC amplifier is given to the PMMC meter.

**iii. Scheme of DC and AC Current measurement**

The current is converted into voltage during shunt arrangements. The variable R is used for multi-ranging facility. The Circuit diagram of AC Current Measurement is shown in Figure 5.26.

**B. Resistance Measurement**

Resistance to voltage convertor consists of a pair of resistors in series. It converts the resistance into DC voltage. The output of DC voltage is given to the amplifier circuit. The block diagram of resistance measurement is shown in Figure 5.27 (a) and (b).

When unknown resistance ($R_x$) is connected to the multi-meter terminals, the internal battery supplies the current through one of the range resistors and the unknown resistance to the ground. The voltage drop ($V_x$) is applied to the input of the bridge amplifier and causes the deflection of the meter. Since the voltage drop across resistance ($R_x$) is directly proportional to its resistance, the meter scale can be calibrated in terms of resistance.

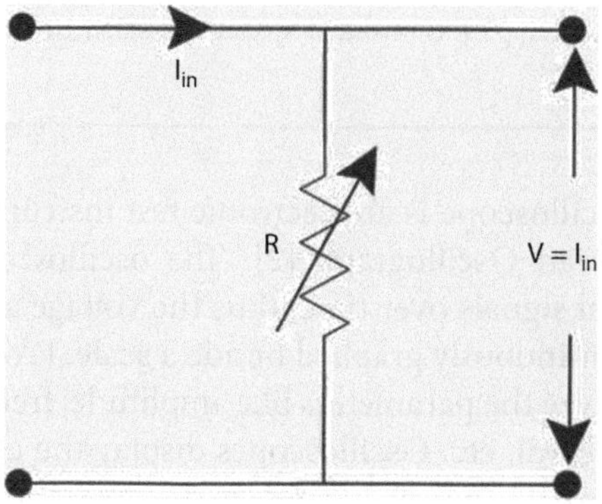


**Figure 5.26** DC and AC voltage measurement.

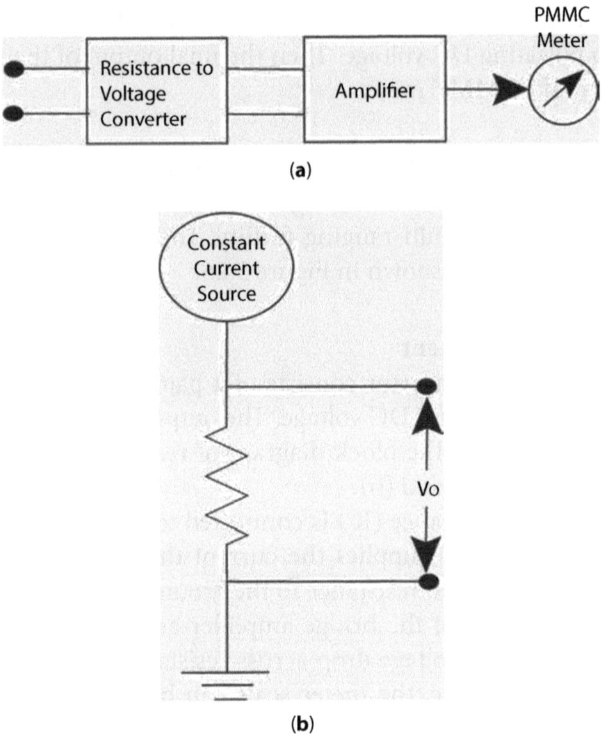


**Figure 5.27** (a) and (b) Resistance measurement.

## 5.2 Cathode Ray Oscilloscope (CRO)

> The cathode ray oscilloscope was invented by German scientist Karl Ferdinand Braun in 1897.

The cathode ray oscilloscope is an electronic test instrument. In the early days, it was called an Oscillograph [2]. The oscilloscope observes the changes in electrical signals over time, thus the voltage and time describe a shape and it is continuously graphed beside a scale. From the waveform, it is possible to analyze the parameters like amplitude, frequency, rise time, distortion, time interval, etc. Oscilloscopes display the change of an electrical signal with respect to time, voltage in Y – axis and time in X - axis, on a calibrated scale.

The oldest form of oscilloscope, still used in some labs today, is called cathode-ray oscilloscope. It produces an image by causing a focused electron beam to travel or sweep in patterns across the face of a Cathode Ray Tube (CRT) [14]. More modern oscilloscopes electronically replicate the action of the CRT using a liquid crystal display similar to those found on notebook computers. The most sophisticated oscilloscopes employ computers to process and display waveforms. These computers can use any type of display including CRT, LCD and gas plasma.

An Oscilloscope display AC or pulsating DC waveform. Low end oscilloscope can display 1hertz and High end oscilloscope can display giga hertz.

Time is displayed from left to right on the horizontal scale. Instantaneous voltage appears on the vertical scale, with positive values going upward and negative values going downward.

In an oscilloscope, the horizontal sweep is measured in seconds per division (Sec/div), milliseconds per division (ms/div), microseconds per division (μs/div) or nanoseconds per division (ns/div). The vertical deflection is measured in Volts per division (V/div), millivolts per division (mV/div) or microvolts per division (μV/div). Virtually all oscilloscopes have adjustable horizontal sweep and vertical deflection settings.

The block diagram of CRT is shown in Figure 5.28. The cathode ray tube acts as a heart of the oscilloscope. In an oscilloscope, the CRT produces the electron beam which is accelerated to a high velocity and

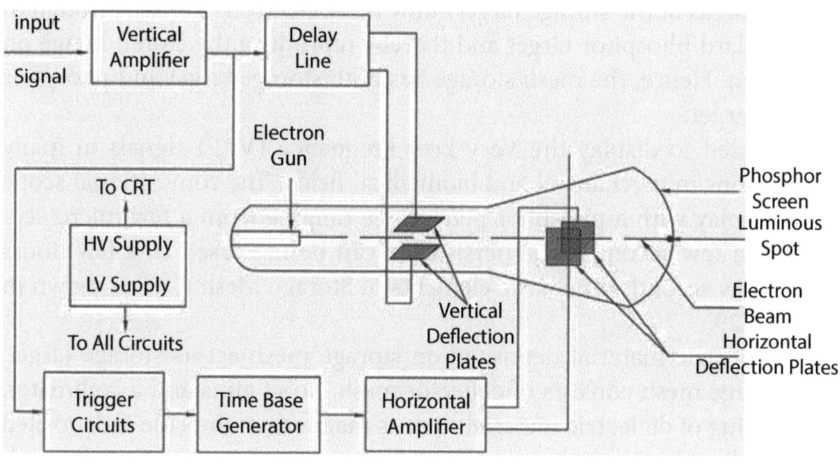


**Figure 5.28** Block diagram of cathode ray oscilloscope.

brings to the focal point on a fluorescent screen [14]. Thus, the screen produces the visible spot where the electron beam strikes with it. By detecting the beam above the screen in reply to the electrical signal, the electrons can act as an electrical pencil of light which produces a light where it strikes.

The low voltage is used for the heater of the electron gun to generate the electron beam. The high voltage is required for the cathode ray tube to speed up the beam. The normal voltage supply is necessary for other control units of the oscilloscope.

The CRT uses vertical and horizontal plates for focusing the electron beam. The horizontal and vertical plates are placed between the electron gun and the screen.

**A. Two storage techniques are used in oscilloscope CRTs**

1. Mesh storage
2. Phosphor storage.

**i. Mesh Storage**

A mesh storage oscilloscope uses the dielectric material deposited on a storage mesh as the storage target. This mesh is placed between the deflection plates and the standard phosphor target in the CRT. The writing beam, which is the focused electron beam of the standard CRT, charges the dielectric material positively where hit [2]. The storage target is then bombarded with low velocity electrons from a flood gun and the positively charged areas of the storage target allow these electrons to pass through to the standard phosphor target and thereby reproduce the stored image on the screen. Hence, the mesh storage has both storage target and phosphor display target.

It is used to display the Very Low Frequency (VLF) signals in many applications in mechanical and biomedical fields. The conventional scope has a display with a phosphor persistence ranging from a few micro seconds to a few seconds. The persistence can be increased to a few hours from a few seconds. The basic elements of Storage Mesh CRT is shown in Figure 5.29.

A dielectric material deposited on storage mesh act as storage target. The storage mesh consists of collector mesh, flood guns and a collimator. The coating of dielectric material such as magnesium fluoride is deposited in a thin layer on the storage mesh. It causes secondary emission. In mesh storage is appropriate for intensity modulation and variable persistence operation.

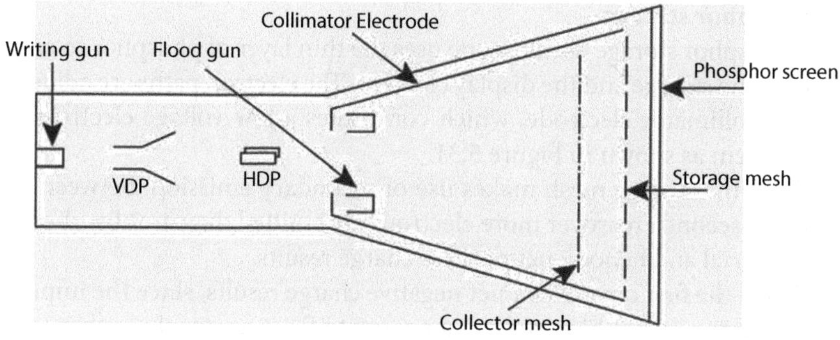


**Figure 5.29** Basic elements of storage mesh CRT.

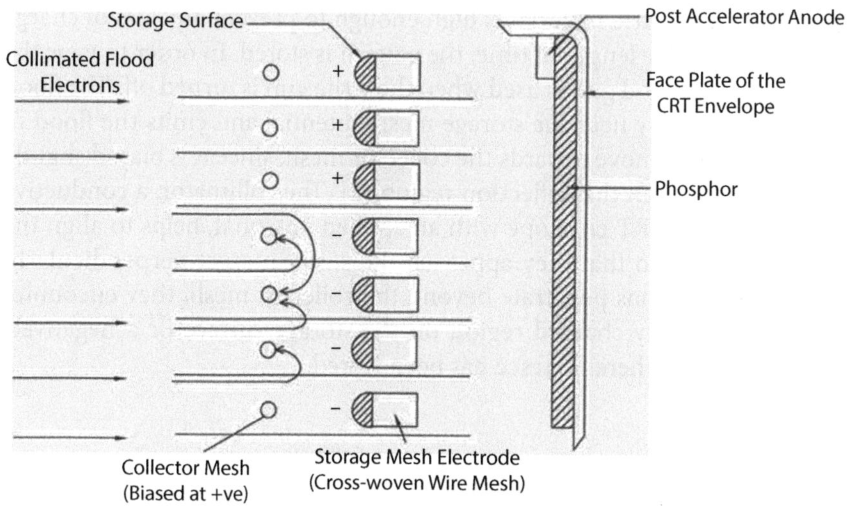


**Figure 5.30** The charge pattern on a mesh storage.

Figure 5.30 shows a display of the stored charge pattern on mesh storage. The storage mesh voltage is +3.3v while potential of storage surface is -10v. To erase the pattern on the storage mesh, a negative voltage is applied to neutralize the stored positive charge. Since the storage mesh makes use of secondary emission, between the first and second crossover more electrons are emitted than are absorbed by the material, and hence a net positive charge results.

**ii. Phosphor storage**

The phosphor storage oscilloscope uses the thin layer of phosphor to serve both as the storage and the display element. The electron paths are adjusted by the collimator electrode, which constitutes a low voltage electrostatic lens system as shown in Figure 5.31.

Since the storage mesh makes use of secondary emission, between the first and second crossover more electrons are emitted than are absorbed by the material and hence a net positive charge results.

Below the first crossover a net negative charge results, since the impinging electrons do not have sufficient energy to force an equal number to be emitted. In order to store a trace, assume that the storage surface is uniformly charged and write gun will hit the storage target [6]. Those areas of the storage surface hit by the deflecting beam lose electrons, which are collected by the collector mesh. Hence, the write beam deflection pattern is traced on the storage surface as a positive charge pattern. Since the insulation of the dielectric material is high enough to prevent any loss of charge for a considerable length of time, the pattern is stored. In order to view the stored trace, a flood gun is used when the write gun is turned off. The flood gun is biased very near the storage mesh potential and emits the flood of electrons which move towards the collector mesh, since it is biased slightly more positive than the deflection region [6]. The collimator, a conductive coating on the CRT envelope with an applied potential, helps to align the flood electrons so that they approach the storage target perpendicularly. When the electrons penetrate beyond the collector mesh, they encounter either a positively charged region on the storage surface or a negatively charged region where no trace has been stored.

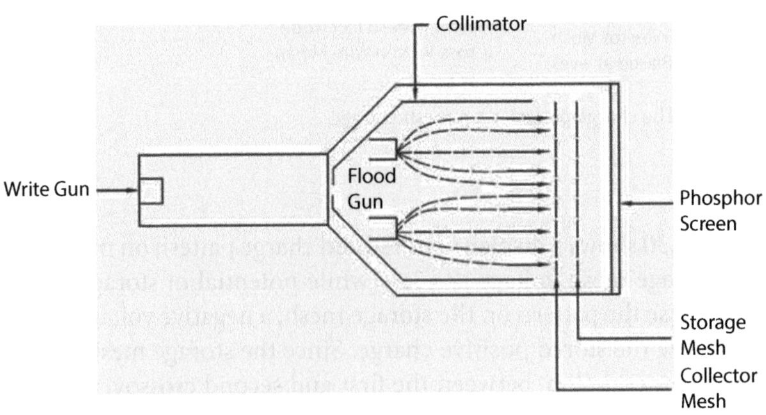


**Figure 5.31** Phosphor storage oscilloscope.

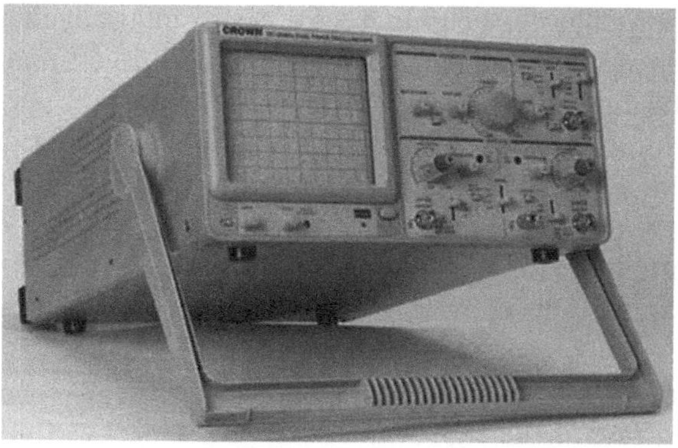

**Figure 5.32** Images of CRO (Courtesy: Cathode Ray Oscilloscope, Product Type: Bio chemical Analysis, Kelvin Labs, Alappuzha, Kerala).

The positively charged areas allow the electrons to pass through to the post accelerator region and the display target phosphor. The negatively charged region repels the flood electrons back to the collector mesh. Thus the charge pattern on the storage surface appears reproduced on the CRT display phosphor just as though it were being traced with a deflected beam. The typical image of CRO is shown in Figure 5.32.

The applications of CRO are as follows

- Voltage measurement
- Current measurement
- Examination of waveform
- Measurement of phase and frequency
- Medicine
- Engineering
- Automotive Industry
- Telecommunication Industry

## 5.3 Digital Storage Oscilloscope

The Digital Storage oscilloscope has been developed for CERN research facility by Walter LeCroy, founder of the company LeCroy. Starting from the 1980s, development of digital oscilloscopes has increased rapidly, which has made them irreplaceable.

A Digital Storage Oscilloscope is an instrument that analyses the signal digitally and stores the data in the electronic digital memory [15]. The digital storage oscilloscope eliminates the disadvantage of an analog oscilloscope. It replaces the unreliable storage method used in analog storage scope with the digital storage with the help of memory.

The maximum frequency measured by digital oscilloscope depends upon two things:

i. Sampling rate
ii. Nature of the converter.

**Sampling Rate**

The sampling rate of the signal must be twice as fast as the highest frequency of the input signal [2]. The sampling rate means rate of analog to digital conversion.

**Converter**

The converter uses the expensive flash whose resolution decreases with the increases of a sampling rate. Because of the sampling rate, the bandwidth and resolution of the oscilloscope are limited. Converter is either analog or digital. The traces in digital oscilloscope are bright, highly defined and displayed within seconds as they are non-stored traces.

The block diagram of digital storage oscilloscope is shown in Figure 5.33. In this method, the waveform is stored in digital memory and retrieved for display on the storage oscilloscope [2].

**Preamplifier**

An analog input signal is fed into the pre amplifier unit. The Preamplifier is used to raise the level of the amplitude of the input signal. Preamplifier signal is fed to an analog to digital convertor and the trigger unit [12].

**ADC**

An analog to digital converter (ADC) is used to change the measured voltages into digital form. Waveforms are taken as a series of samples. The output of the pre-amplifier is sampled by the ADC at regular intervals. The

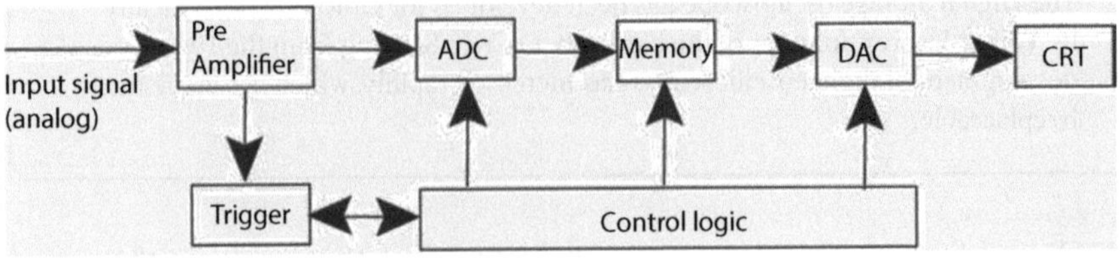


**Figure 5.33** Block diagram of digital storage oscilloscope.

digital output provided by the ADC is stored in memory at consecutive locations.

**Memory**

The digital storage oscilloscope (DSO) also has two types of memory acquisition and display memory. A real time digital storage oscilloscope acquires a long data record in an acquisition memory and processes the data of the long data record to search for predetermined events.

**DAC**

The digital to analog convertor produces the analog signal to be displayed in CRT. Any further changes in the applied input re-triggers the oscilloscope which causes the memory to reset. That means the memory is overwritten with the new upcoming data unless the system is in HOLD mode. By the application of HOLD mode, the signal traces can be analyzed by the user according to the user's choice. The size of the memory unit determines the number of samples stored in it. One can alter the length of recording by changing the sampling frequency of ADC. In digital oscilloscope, multiple input channels are used. But, the memory available to each channel gets reduced. This is done through multiplexing.

The digital storage oscilloscope has different modes:

1. Roll mode
2. Store mode or Refresh Mode
3. Hold or save mode
4. Equivalent Time Mode

**1. Roll Mode**

Roll mode is the most basic mode of operation which is similar to that of a general purpose CRO. When an input is applied, trace is displayed on the screen. A user can use this mode to keep an eye on the waveform and its various characteristics.

**2. Store mode**

Store mode is otherwise called refresh mode. In this mode, the input initiates the trigger circuit. This initiates the memory write cycle. The digital data is transferred to the memory when the memory is full, the write cycle stops. Here, using DAC, the memory data is converted to analog and then displayed on the screen when the next trigger across the memory is refreshed.

**3. Hold**

Hold is otherwise called save mode. When a new sweep signal is generated by time base generator, the old contents get overwritten by the new one. If a particular signal is to be stored then by processing hold or save button, overwriting can be stopped and the previously saved signal is locked.

**4. Equivalent Time Mode**

It basically arranges the ADC to work on limited sweeps, so as to allow particular signals to sync properly and give out a clearer trace.

**Waveform Reconstruction**

The input is sampled according to sampling theory; aliasing effect can still occur. This is because the output is present as a series of dots that correspond to the sampled value. Digital storage oscilloscope uses interpolation technique for visualization of final waves. Interpolation is a technique that creates new data points with the help of a discrete set of known data points.

Interpolation is basically of two types:

- Linear interpolation
- Sinusoidal interpolation

The interpolation techniques are listed in Table 5.2.

The main advantage of the digital oscilloscope is that it can display visual as well as numerical values by analyzing the stored traces as shown in Figure 5.34.

Digital oscilloscopes usually analyze waveforms and provide numerical values as well as visual displays. These values typically include averages, maxima and minima, root mean square (RMS) and frequencies. They may be used to capture transient signals when operated in a single sweep mode, without the brightness and writing speed limitations of an oscilloscope. The image of Digital Storage Oscilloscope is shown in Figure 5.35.

The advantages of digital storage oscilloscopes are as follows:

1. The storage time is infinite
2. It can be easily operated and has more capability
3. Digital storage oscilloscope allows flexible display property
4. The cursor measurement is possible
5. The X-Y plots, B-H curve, P-V diagrams can be displayed

**Table 5.2** Methods of basic interpolation techniques.

| S. No. | Methods | Explanation | Waveform |
|---|---|---|---|
| 1 | Without Interpolation | | 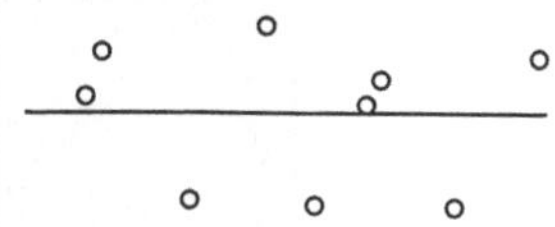 |
| 2 | Linear Interpolation | In linear interpolation, the dots are connected together by a straight line. It can be used in pulsed or square wave generation but not in case of sinusoidal waves. | 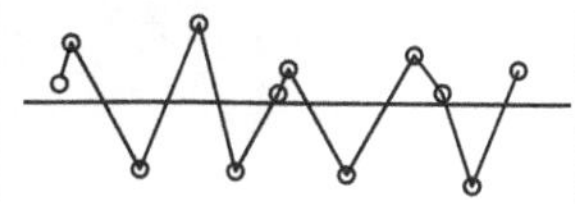 |
| 3 | Sinusoidal Interpolation | In sinusoidal interpolation, the dots are connected so as to form sinusoidal waveform. But it is not suitable for square or pulse waves. | 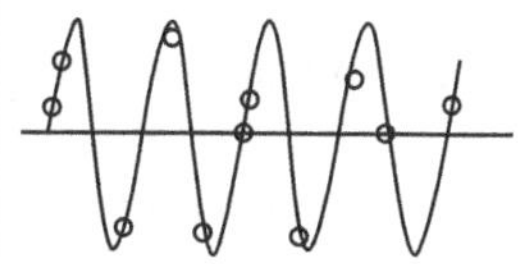 |

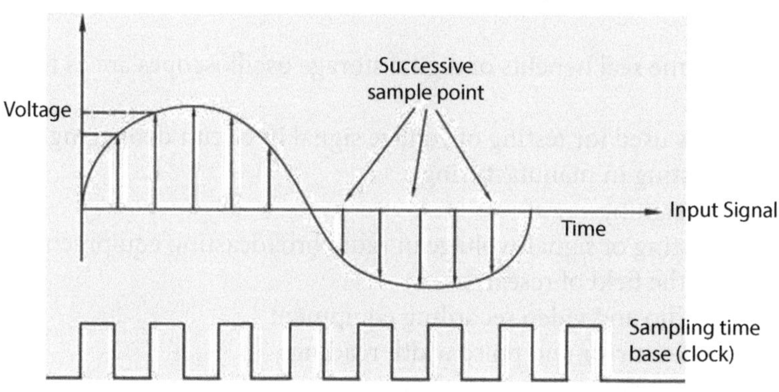


**Figure 5.34** Waveform of digital storage oscilloscope.

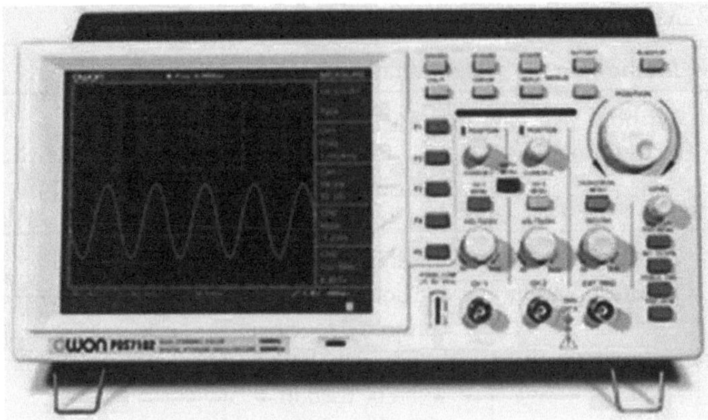

**Figure 5.35** Digital storage oscilloscope (Courtesy: Digital Storage Oscilloscope, Product Type: DSO 60MHz (Color LCD Display-AE 482): Ambala Electronics Instruments, Haryana).

The applications of digital storage oscilloscopes are as follows:

1. It is used in audio and video recording
2. It is used in radio broadcasting for signal testing
3. In circuit debugging, it is used for testing of the voltage of the signal
4. The measurement of various AC parameters such as current, voltages, etc.
5. It is used to measure slow moving parameters such as temperature of the day

Some of the real benefits of digital storage oscilloscopes are as follows:

- It is used for testing of voltage signal in circuit debugging
- Testing in manufacturing
- Designing
- Testing of signals voltage in radio broadcasting equipment
- In the field of research
- Audio and video recording equipment
- Take cursor and pulse width readings
- Measuring rise time and propagation delay
- Implementing math functions like subtraction and addition
- The DSO can also measure components' minor variations in operations and alert the technician of the **need** for fine-tuning or replacement

- To prevent erroneous replacement of parts, the DSO also helps technicians identify the parts that are still working

## 5.4 Static and Dynamic Characteristics of Measurements

### 5.4.1 Static Characteristics

The characteristics of various quantity measurements by instruments that do not vary with respect to time are called static characteristics [1, 2, 8]. These quantities can vary slowly with respect to time.

i. Accuracy
ii. Precision
iii. Sensitivity
iv. Resolution
v. Threshold
vi. Linearity
vii. Drift
viii. Reproducibility
ix. Stability
x. Hysteresis
xi. Resolution
xii. Threshold

**i. Accuracy**
The algebraic difference between the indicated value of an instrument and the true value is known as accuracy. Mathematically, it can be represented as

$$\textbf{Accuracy} = \textbf{Ai} - \textbf{At}$$

Where,

Indicated Value = Ai
True Value = At

Also it can be expressed in

a. Accuracy as "percentage of full scale reading"
b. Percentage of True Value

c. Point Accuracy
d. Accuracy as percentage of Scale Span

**a. Accuracy as "percentage of full scale reading"**

- The accuracy can be expressed as a percentage of full scale reading
- The accuracy of an instrument having full scale reading of 50 units may be expressed as +/- of full scale reading

**b. Percentage of True Value**
It is specified in terms of the true value of quantity being measured. For example, it can be specified as +/- 0.1% of true value.

**c. Point Accuracy**
It is specified at only one particular point of scale. It does not give any information about the accuracy at any other point on the scale.

**d. Accuracy as percentage of Scale Span**
When an instrument is in uniform scale, its accuracy may be expressed in terms of scale range.

$$\text{Span} = (a_{\max} - a_{\min})$$

$a_{max}$ is the maximum point for which scale is calibrated
$a_{min}$ is the lowest reading on scale

**ii. Precision**
It is measure of reproducibility or degree of agreement within a group of measurements.

**Example:**
If an instrument indicates the same value repeatedly when it is used to measure the same quantity under the same circumstances for any number of times, then the instrument has high precision.

**iii. Sensitivity**
It is defined as the ratio of the changes in the output of an instrument to a change in the value of the quantity to be measured [8]. Mathematically it is expressed as

$$\text{Sensitivity} = \frac{\text{Infinite change in output}}{\text{Infinite change in input}}$$

$$\text{Sensitivity} = \frac{\Delta_{qo}}{\Delta_{qi}}$$

The term sensitivity signifies the smallest change in the measurable input that is required for an instrument to respond as shown in Figure 5.36 (a) & (b).

- If the calibration curve is linear, then the sensitivity of the instrument will be a constant and it is equal to the slope of the calibration curve
- If the calibration curve is non-linear, then the sensitivity of the instrument will not be a constant and it will vary with respect to the input

Measuring instruments that have smaller scale parts are more sensitive. Sensitive instruments need not necessarily be accurate.

**Ex : Galvanometer**

A Galvonometer is an electromechanical instrument used for detecting and indicating an electric current. A Galvonometer works as an actuator by producing rotary deflection in response to electric current flowing through a coil in a constant magnetic field. The typical Galvonometr name plate and actual meter is shown in Figure 5.37 and Figure 5.38, respectively.

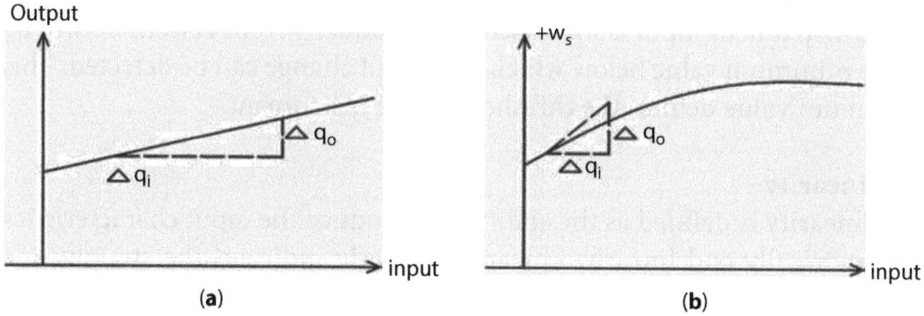


**Figure 5.36** (a) and (b) Waveform of input vs. output.

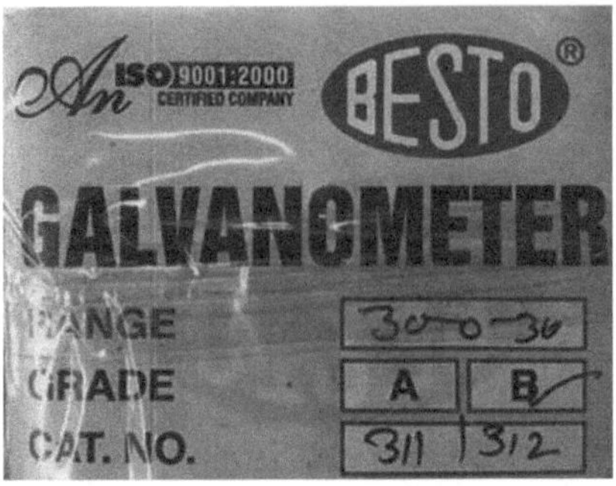


**Figure 5.37** Galvonometer name plate (Courtesy: Galvanometer Analog Besto (311/312 A)).

**Figure 5.38** Galvonometer (Courtesy: Galvanometer - Analog - BESTO (311/312 A)).

### iv. Resolution

If the output of an instrument change only when there is a specific increment of the input, then the increment of the input is called Resolution.

### v. Threshold

If the instrument input is increased very gradually from zero there will be some minimum value below which no output change can be detected. This minimum value defines the threshold of the instrument.

### vi. Linearity

The linearity is defined as the ability to reproduce the input characteristics symmetrically and linearly. Linearity properly indicates the straight line nature of the calibration curve shown in Figure 5.39.

$$\text{Linearity } (\%) = \frac{\text{Maximum deviation of output from idealized straightline}}{\text{Full Scale Deflection}} * 100$$

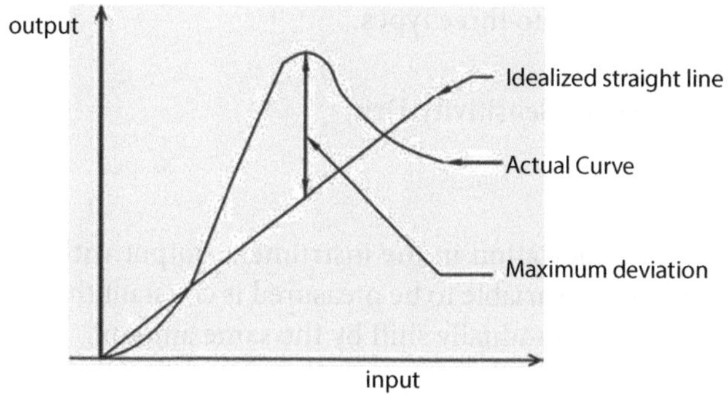


**Figure 5.39** Input vs. output of linearity.

It is classified based on methods of measurement:

a. Direct measuirng Instruments
b. Indirect measuring instruments

Direct Measuring instruments can be further classified as

- Graduated Instruments (Steel Rule, Vernier Caliper)
- Non-Graduated Instruments (Surface Gauge)

**➢ Graduated Instruments**

Graduated measuring device include a set of markings on a linear [16]. The most basic of the graduated measuring device is the rule. It is used to measure linear dimensions. It includes Vernier Calllipers, Vernier height gauges, Vernier depth gauges, Micrometers, Dial indicators, etc [17].

**➢ Non-graduated Instruments**

Non graduated measuring device is used to make comparisons between dimensions.

It includes Callipers, Wire Gauges, Screw Pitch Gauges, Radius Gauges, Thickness Gauges, Slip Gauges, etc.

**vii. Drift**

It is defined as the gradual shift in the indication over a period of time wherein the input variable does not change. Drift may be caused because of environmental factors like stray electric fields, stray magnetic fields, changes in temperature, mechanical vibrations etc.

It can be classified into three types:

- Zero Drift
- Span Drift or Sensitivity Drift
- Zonal Drift

➢ **Zero drift**

It is defined as the deviation in the instrument output with time from its zero value. When the variable to be measured is constant the whole instrument calibration may gradually shift by the same amount.

**Example:**

If the whole calibration gradually shifts due to slippage, permanent set or due to undue warming up of electronic tube circuits, zero drift sets in.

➢ **Span Drift or Sensitivity Drift**

If there is proportional change in the indication all along the upward scale, the drift is called span drift or sensitivity drift as shown in Figure 5.40 (a) & (b).

➢ **Zonal Drift**

When the drift occurs only over a portion of the full scale or span of an instrument, it is called zonal drift.

**viii. Reproducibility**

The degree of closeness which a given value may be repeatedly measured. It may be specified in terms of units for a given period of time. The reproducibility is shown in Figure 5.41.

**ix. Stability**

The ability of an instrument to retain its performance throughout its specified storage life and operating life is called stability. The stability measurement instruments shown in Figure 5.42.

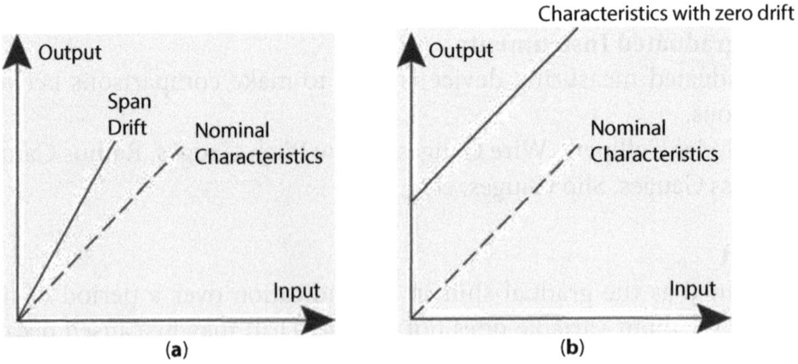


**Figure 5.40** (a) Span drift. (b) Zero drift.

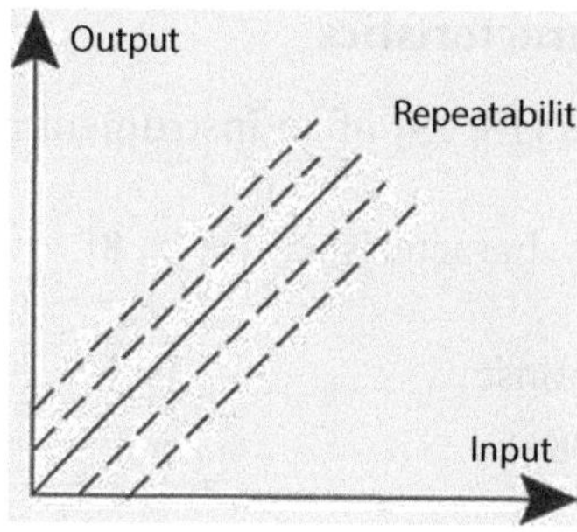


**Figure 5.41** Reproducibility.

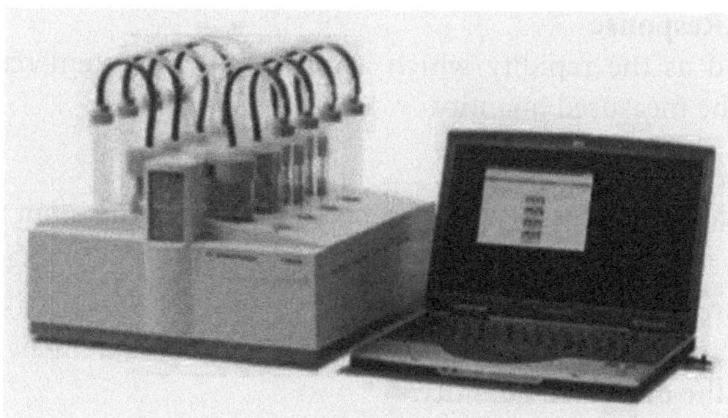

**Figure 5.42** Image of stability measuring instruments (Courtesy: Stability Measuring Instruments, Metrohm India Ltd, Chennai).

### x. Hysteresis

If the input to the instrument is increased from a negative value, the output also increases.

### xi. Resolution

If the input is slowly increased from some arbitrary input value, it will again be found that output does not change at all until a certain increment is exceeded. This increment is called resolution.

### xii. Threshold

If the instrument input is increased very gradually from zero there will be some minimum value below which no output change can be detected. This minimum value defines the threshold of the instrument.

### 5.4.2 Dynamic Characteristics

Dynamic characteristics [18, 19] of an instrument measure the quantities which vary with time.

The various dynamic characteristics are [2, 8]

i. Speed of response
ii. Measuring lag
iii. Fidelity
iv. Dynamic error
v. Rise Time
vi. Settling Time

**i. Speed of Response**
It is defined as the rapidity which a measurement system responds to change in the measured quantity.

**Example:**
It indicates how fast the sensor reacts to changes in the input variable.

**ii. Measuring lag**
It is the retardation or delay in the response of a measurement system to changes in the measured quantity.

The measuring lags are two types

a. Retardation type
b. Time Delay Lag

**a) Retardation type**
In this case, the response of the measurement system begins immediately after the change in measured quantity has occurred.

**b) Time delay lag**
In this case, the response of the measurement system begins after a dead time after the application of the input.

**iii. Fidelity**
It is defined as the degree to which a measurement system indicates changes in the measured quantity without dynamic error.

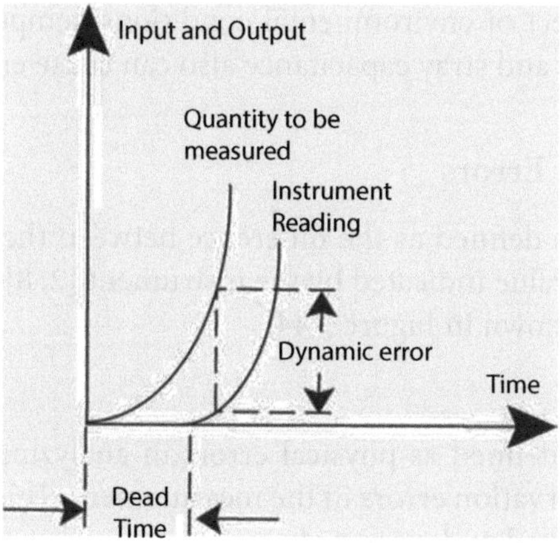


**Figure 5.43** Dynamic error.

**iv. Dynamic error**

The difference between the true value of the quantity changing with time and the value indicated by the measurement system if there is no static error is called dynamic error or measurement error. The dynamic error is shown in Figure 5.43.

**v. Rise time**

The length of time it takes the output to reach 90% of full response when a step is applied to the input.

**vi. Settling time**

The time it takes from the application of the input step until the output has settled within the specific band of final value.

## 5.5 Measurement of Errors

The faulty design of the instrument which directly leads to serious measurement errors [20]. Insufficient knowledge of the quantity to be measured and design conditions can cause errors.

If the maintenance is not carried out properly or periodically, errors may occur.

i. Unskilled Operator
ii. Loading effect

iii. An effect of environmental conditions, temperature conditions and stray capacitance also can cause errors

### 5.5.1 Types of Errors

The static error is defined as the difference between the true value of the variable and the value indicated by the instrument [2, 8]. The flowchart of types of error is shown in Figure 5.44.

**A. Gross errors**

Gross errors are defined as physical errors in analyzing the values from apparatus or observation errors in the measurement. This is a human error that can be estimated and corrected.

i. It is mainly caused by lack of experience

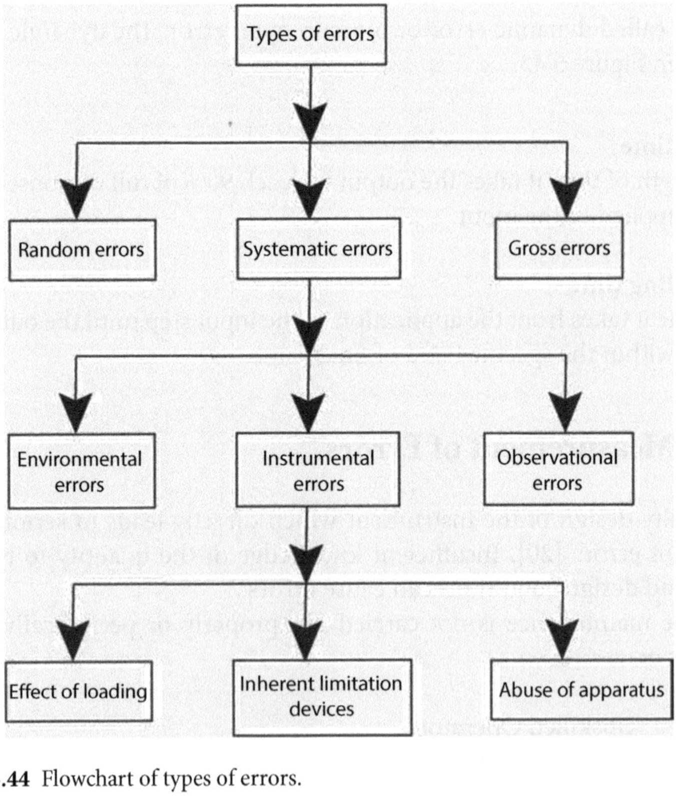


**Figure 5.44** Flowchart of types of errors.

ii. These cover human mistakes in readings, recordings and calculating the results
iii. Due to incorrect adjustment of instrument

This error can be minimized by the following ways:

i. By avoiding careless mistakes

ii. Multiple readings must be taken and preferably by different persons

**B. Systematic errors**

These are due to the shortcoming of the instrument and the characteristics of the material used in the instruments, such as defective parts, aging effects, environmental effects etc., These types of systematic errors are generally categorized into three types:

i. Instrumental Errors
ii. Environmental Errors
iii. Observational Errors

**i. Instrumental errors**

The instrumental error refers of the combined accuracy and precision of a measuring instrument.

Instrumental errors are categorized into three types:

1. Due to inherent shortcoming in the instrument
2. Due to misuse of Instrument
3. Effect of Loading

**1. Due to inherent shortcoming in the instrument**

This error is integral in devices due to their features, namely mechanical arrangement. This may happen due to the instrument operation, and this error will make readings either very low or very high.

**Example:** If the spring used in permanent magnet instrument has become weak then instrument will always read high. Error is because of friction, hysteresis or even gear backlash.

**2. Misuse of Instruments**

The error in the instrument happens due to the machinist's fault. A superior device used in an undesired way can provide the misleading results.

**3. Effect of Loading**

Loading effects are due to an improper way of using the instrument, causing the serious errors. For instance, as the voltmeter is associated with the high resistance circuit which will give a false reading, as well as after it is allied to the low-resistance circuit, this circuit will give the reliable reading and then the voltmeter has the effect of loading on the circuit.

**ii. Environmental Errors**

Environmental errors will happen due to environment conditions of the measuring instruments. These types of errors mostly happen due to the temperature result, force, moisture, dirt, vibration, electrostatic field or magnetic field.

This error can be minimized by the following ways:

i. The arrangements which keep the surrounding conditions as constant. This includes the use of air conditioning, temperature control enclosures, etc.
ii. It reduces the effect of dust, humidity.
iii. The effect of external fields can be minimized by using the magnetic or electrostatic shields.

**iii. Observational errors**

There are many sources of observational errors, such as parallax error while reading a meter, wrong scale selection and the habits of individual observes, etc.

**C. Random errors**

Random errors are errors in measurement that lead to measurable values being inconsistent when repeated measurements of quantity are taken. Random errors are unpredictable and can't be replicated by repeating the experiment.

This can be reduced by

i. Using an average measurement from a set of measurements
ii. Increasing sample size

## 5.6 Transducer

A transducer is a device which transform a non-electrical physical quantity (temperature, sound or light) into an electrical signal (voltage, current) [2, 21].

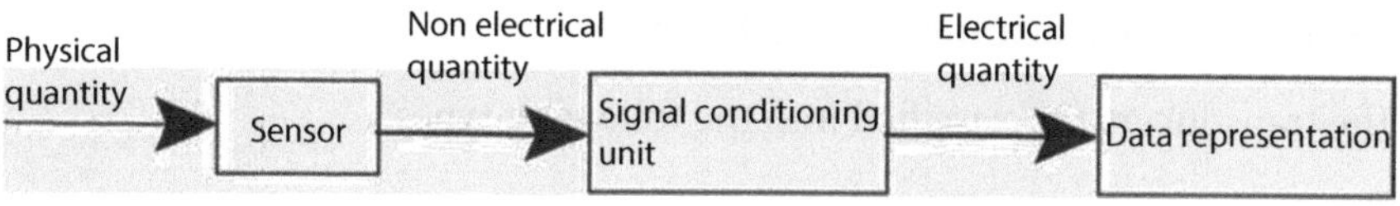


**Figure 5.45** Block diagram of measuring system.

The typical block diagram of a simple measuring system is shown in Figure 5.45.

It consists of three basic elements are as follows,

- Sensor
- Signal Conditioning Unit
- Data Representing Device

**Sensor**

Sensor is a device that produces an output signal for the purpose of sensing a physical phenomenon [22]. A sensor is a device that receives and responds to a signal. Sensor is used to detect changes in any physical quantity like Temperature, Speed, light, vibration etc. The input quantity can be detected by a sensor and reflected as changes in output quantity.

**Signal Conditioning Unit**

It is used to convert the physical quantity (non-electrical output) measured by the sensor to an electrical quantity.

Some of the known signal conditioning units is

- Analog to Digital Converters
- Amplifiers
- Filters
- Rectifiers
- Modulators

**Data Representation Device**

A data representation device is used to present the measured output to the observer. This can be anything like

- A Scale
- An LCD Display
- A Signal Recorder

### 5.6.1 Classification of Transducers

The transducers are classified into the following types:

1. Based on the physical phenomenon
   - Primary transducer
   - Secondary transducer
2. Based on source of energy
   - Active transducer
   - Passive transducer
3. Based on the type of output
   - Analog transducer
   - Digital transducer
4. Based on the transduction phenomenon
   - Transducer
   - Inverse transducer
5. Based on the electrical phenomenon
   - Resistive transducer
   - Capacitive transducer
   - Inductive transducer
   - Photoelectric transducer
   - Photovoltaic transducer
6. Based on the non-electrical phenomenon
   - Linear displacement
   - Rotary displacement

**A. Based on the physical phenomenon**

Based on the physical phenomenon, transducers are classified into:

- Primary transducer
- Secondary transducer

**i. Primary Transducer**

The transducer consists of mechanical as well as electrical devices. The mechanical devices of the transducer change the physical input quantities into a mechanical signal. This mechanical device is known as the primary transducer.

**Example:** Bourdon Tube, Load cell, Bellows

**ii. Secondary Transducer**

The secondary transducer converts the mechanical signal into an electrical signal. The magnitude of the output signal depends on the input mechanical signal.

**Example:** LVDT, Strain Gauge

**iii. Example of Primary and Secondary Transducer**

**a. Bourdon Tube**

The Bourdon's Tube acts as a primary transducer. It detects the pressure and converts into a displacement from its free end. The displacement of the free ends moves the core of the linear variable displacement transformer. The movement of the core induces the output voltage which is directly proportional to the displacement of the tube free end.

The pressure is converted into a displacement. The Bourdon's Tube is the primary transducer.

**b. LVDT**

The Displacement converted into the voltage by LVDT is called the secondary transducer [23]. A block diagram of LVDT is shown in Figure 5.46.

The comparison of primary and secondary transducers are listed in Table 5.3.

**B. Based on source of energy**

Based on the source of energy, transducers are classified into:

- Active transducer
- Passive transducer

**i. Active transducers**

The active transducer is a device which converts the given non-electrical energy into electrical energy by itself [2].

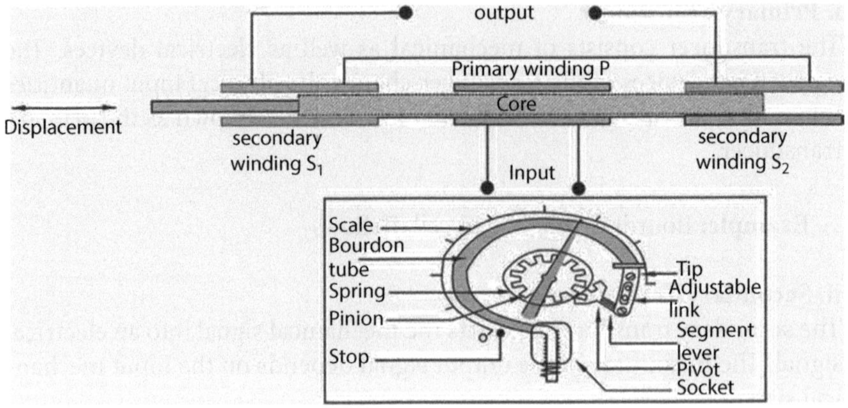


**Figure 5.46** Block diagram of LVDT.

**Table 5.3** Comparison of primary and secondary transducer.

| S. No. | Comparison | Primary transducer | Secondary transducer |
|---|---|---|---|
| 1 | Device | It is a mechanical device | It is an electrical device |
| 2 | What is | The force is detected by the column in first so it is called Primary Transducer | Output of Primary transducer converts unto useful output signal; is known as secondary transducer |
| 3 | Examples | Load Cell, Bourdon Tube | LVDT, Strain Gauge |

**Example:**
Thermocouple, photovoltaic cell, etc.

## Thermocouple

The German physicist Thomas Johann Seeback discovered the Thermocouple in 1821.

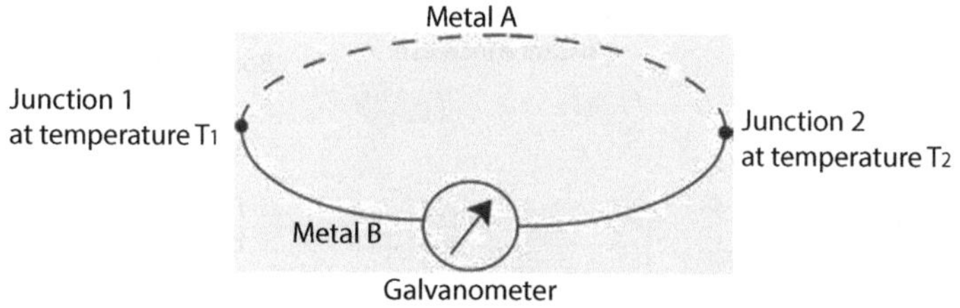


**Figure 5.47** Thermocouple.

The working principle of a thermocouple is based on the Seeback effect. An EMF is generated across the two dissimilar metal junctions when the junction is heated. A thermocouple consists of two pieces of dissimilar metals with their ends joined together When heat is applied to the junction, voltage in the range of milli-Volts (mV) is generated. A thermocouple is therefore said to be self-powered. The typical thermocouple is shown in Figure 5.47.

The voltage generated at each junction depends on junction temperature. If temperature $T_1$ is higher than $T_2$, then the voltage generated at Junction 1 is higher than that at Junction 2. In the above circuit, the loop current shown on the galvanometer depends on the relative magnitude of the voltages at the two junctions.

### ii. Passive transducers

A passive transducer is a device which converts the given non-electrical energy into electrical energy by external force [8].

**Example:**
Resistance strain gauge, LVDT, etc.

### Strain gauges

> In 1843 the English physicist Sir Charles Wheatstone (1802-1875) found a bridge circuit for measuring electrical resistances. In this bridge circuit, known today as the Wheatstone bridge circuit, unknown resistances are compared with well-defined resistances.

It has simple sensors that can be used to measure forces. It consists of a piece of conducting material that changes resistance as it is stretched in a given direction. The Strain Gauge is shown in Figure 5.48.

The change in resistance of the strain gauge is very small. The Wheatstone bridge measures small imbalances in the resistances. Here, it is comparing

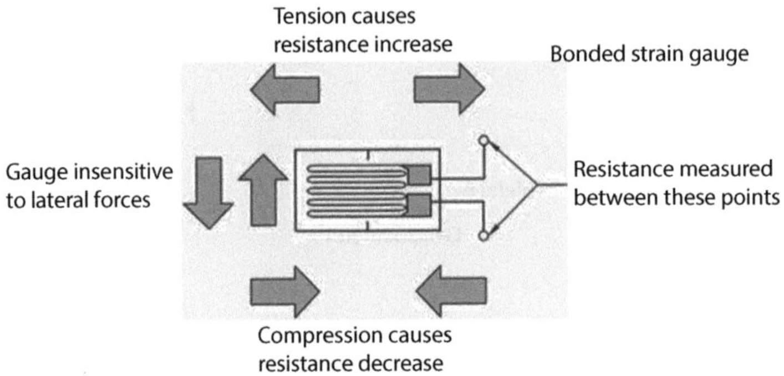


**Figure 5.48** Strain gauge.

the strain gauge resistance to *R*, which has a resistance equal to the resistance of the unstretched strain gauge. The other two resistors should have similar values. The comparisons of active and passive transducers are listed in Table 5.4.

**C. Based on the type of output the classification of transducers are made**
Based on the type of output, transducers are classified into

i. Analog transducer
ii. Digital transducer

**i. Analog Transducer**
The analog transducer changes the input quantity into a continuous function. The strain gauge, LVDT, thermocouple, thermistors are examples of the analog transducer [8].The comparison of analog and digital transducer are listed in Table 5.6.

**Example:** Thermistor

> Michael Faraday; an English scientist, first discovered the concept of thermistors in 1833 while reporting on the semiconductor behavior of silver sulfide. Through his research, he noticed that the silver sulfides resistance decreased as the temperature increased. This discovery would later lead to the commercial production of thermistors in the 1930s when Samuel Ruben invented the first commercial thermistor.

The thermistor is a combination of the words thermal and resistor. A thermistor is a temperature sensing element composed of semiconductor

**Table 5.4** Comparison of active and passive transducer.

| S. No. | Comparison | Active transducer | Passive transducer |
|---|---|---|---|
| 1 | What is | The transducer which generate the output in the form of voltage or current, without any external energy source is known as active transducer | The passive transducer means the transducer whose internal parameters like capacitance, resistance and inductance changes because of the input signal |
| 2 | Additional Energy Source | Not Required | Required |
| 3 | Working Principle | Draw energy from the measure and source | It take power from the external source which changes the physical properties of transducer |
| 4 | Design | Simple | Complicated |
| 5 | Output Signal | It produces from the signal to be measured | Output signal obtains by receiving the signal from the external power source |
| 6 | Examples | Tachogenerator, Thermocouple, Photovoltaic cell, etc. | Thermistor, Differential Transformer, Photomultiplier tube, Photovoltaic cell |

material that exhibits a large change in resistance in response to small change in temperature. Thermistor usually have negative temperature coefficients, the resistance of the thermistor decreases as the temperature increases. Thermistor are available in various shape like bead, probe, disc, rod etc.

**Types of Thermistor**

The thermistor is classified into two types, the negative temperature coefficient and positive temperature coefficient.

**1. Negative Temperature Coefficient Thermistor**
In this type of thermistor the temperature increases with the decrease of the resistance. The resistance of the negative temperature coefficient thermistor is very large due to which it detects a small variation in temperature.

**2. Positive Temperature Coefficient Thermistor**
The resistance of the thermistor increases with the increases in temperature.

The thermistor is made up of sintered mixture of metallic oxides like manganese, cobalt, nickel, cobalt, copper, iron, uranium, etc. It is available in the form of the bead, rod and disc. The different types of the thermistor are shown in Table 5.5.

**ii. Digital Transducer**
Digital transducers convert the input quantity into the digital signal or in the form of the pulse. The digital signals work either on high or low [2].

**Example:** Shaft Encoders

**D. Based on the transduction phenomenon**
Based on the transduction phenomenon, transducers are classified into

- Transducer
- Inverse Transducer

**i. Transducer**
The transducer is a type of device that converts the physical quantities like pressure, light, displacement into an electrical quantity. The process of conversion of quantities is known as transduction. A block diagram of a transducer is shown in Figure 5.49.

**Examples:** The thermocouple converts the temperature into voltage and the LVDT is used for measuring the displacement.

**ii. Inverse Transducer**
The inverse transducer is the transducer which converts the electrical quantity into a non-electrical quantity. In other words, the transducer is a kind of actuator which has an electrical input and the non-electrical output [2]. The comparison of transducer and inverse transducer are listed in Table 5.7.

**Examples:**
The analog ammeter and voltmeter converts the current into a displacement. The oscilloscope is used for converting the electrical signal into a physical displacement.

**Table 5.5** The different types of thermistor.

| S. No. | Type | Diagram | Description |
|---|---|---|---|
| 1 | Bead | 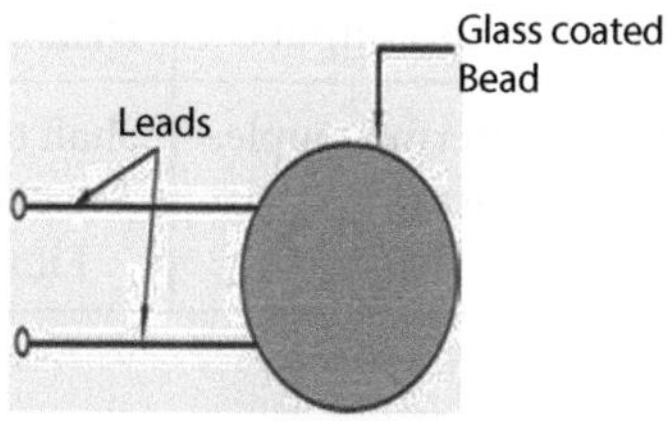  | The bead form of the thermistor is smallest in shape, and it is enclosed inside the solid glass rod to form probes. |
| 2 | Disc | 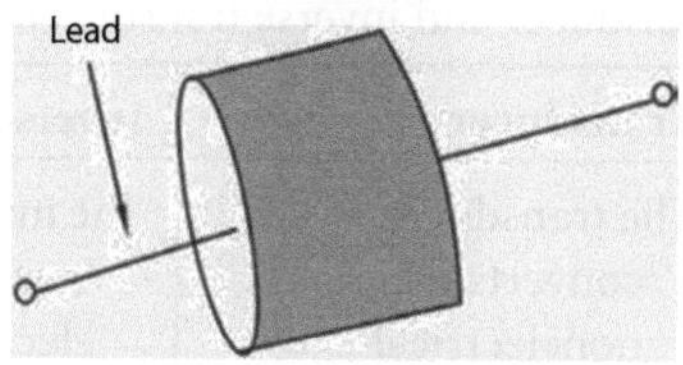  | The disc shape is made by pressing material under high pressure with diameter range from 2.5 mm to 25mm. |
| 3 | Rod | 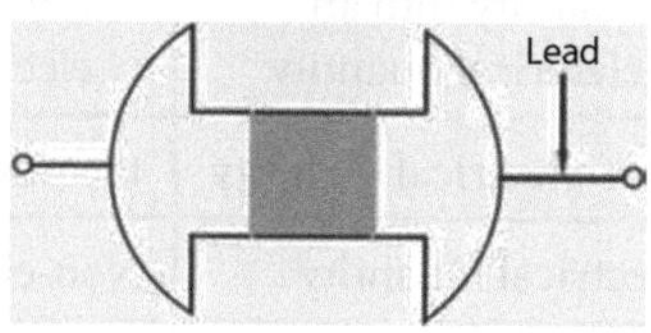  | A type of thermistor that has high resistance, long time constant, and moderate power dissipation; it is extruded as a long vertical rod 0.250-2.0 inches (0.63-5.1 centimeters) long and 0.050-0.110 inch (0.13-0.28 centimeter) in diameter, of oxide-binder mix and sintered; ends are coated with conducting paste and leads are wrapped on the coated area. |

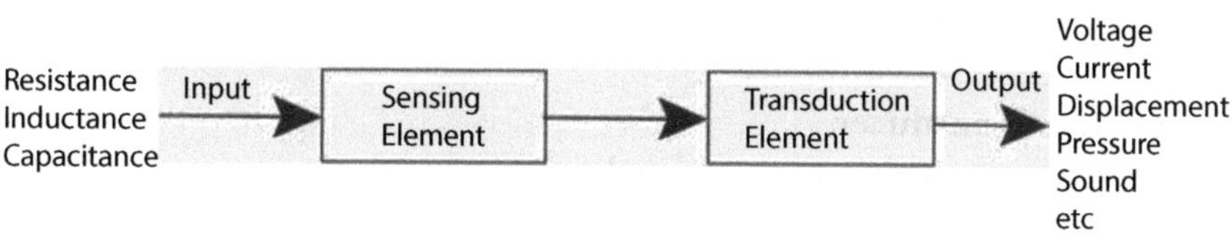


**Figure 5.49** Block diagram of transducer.

Table 5.6 Comparison analog transducer and digital transducer.

| S. No. | Comparison | Analog transducer | Digital transducer |
|---|---|---|---|
| 1 | Output Signal | Continuous Output | Discrete Output. |
| 2 | Example | LVDT, Thermocouple, Thermistor | Shaft encoders, Digital tachometers, and Limit switches. |

Table 5.7 Comparison of transducer and inverse transducer.

| S. No. | Comparison | Transducer | Inverse transducer |
|---|---|---|---|
| 1 | Definition | The transducer converts the nonelectrical quantity into an electrical quantity | The inverse transducer converts the electrical quantity into the non-electrical quantity. |
| 2 | Input | Non-electrical quantity | Electrical quantity |
| 3 | Output | Electrical Quantity | Non-electrical quantity |
| 4 | Example | Photoconductive Transducer Thermocouple, Pressure Gauge, Strain Gauge | Piezoelectric Transducer, Current carrying conductor placed in an magnetic field |

**E. Based on the electrical phenomenon**

Based on the electrical phenomenon, transducers are classified into

- Resistive transducer
- Capacitive transducer
- Inductive transducer
- Photoelectric transducer
- Photovoltaic transducer

**i. Resistive Transducer**

The resistive transducers are also known as resistive sensors or variable resistance transducers [23]. The transducer whose resistance varies because of the environmental effects such type of transducer is known as the resistive transducer. The change in resistance is measured by the ac or dc measuring device. The resistive transducer is used for the measuring the physical quantites like temperature, displacement etc.

**Working of Resistive Transducer**

This is the most frequently used transducer to calculate pressure, temperature, force, displacement, vibrations, etc. These transducers work on the principle of the length of a conductor which is directly proportional to the conductor's resistance and it is inversely proportional to the conductor's area. So, the denominated length of the conductor is 'L', the area is 'A', resistance is 'R' and the resistivity is 'ρ'. It is stable for every material which is used in conductor construction.

$$R = \rho L/A$$

Where,

R is the resistance of the conductor.
A is the side view part of the conductor.
L is the conductor's length.
ρ is the resistivity of the conductor.

The transducer's resistance can be changed because of the exterior environmental factors as well as the conductor's physical properties. A primary transducer is used to change the physical quantity to the mechanical signal whereas a secondary transducer is used to convert a mechanical signal into an electrical signal. The major types of resistive transducer include potentiometers, resistive position transducers, resistive pressure transducers, thermistors, strain gauges and LDR.

**ii. Capacitive Transducer**

The capacitive transducer works on the fundamental principle of electrical capacitance. The capacitance 'C' of a system depends on the dielectric medium used and properties of a capacitive system [23].

**Working Principle**

The working principle of a capacitive transducer is variable capacitance. As per its structure, these are having two parallel metal pates which are maintaining the distance between them. In between them, dielectric medium (Air) can be filled. So, the distance between these two metal plates and positions of the plates can change the capacitance. So, variable capacitance is the principle of these transducers. The basic difference between the normal capacitors and capacitive transducers is, the capacitor plates are constant in normal capacitors wherein these transducers, capacitor plates are the movable condition. The capacitor transducer is shown in Figure 5.50.

The capacitance of the variable capacitor can be measured by

$$C = \frac{\varepsilon_o \varepsilon_r^A}{d}$$

C is the capacitance of the variable capacitance
$\varepsilon_o$ is the permittivity of free space
$\varepsilon_r$ is the relative permittivity
A is area of the plates
d is distance between the plates

So according to the formula, the variable capacitance value is dependent on four important parameters. They are the distance between the plates of the variable capacitor, occupying area of the plates, permittivity of the free space, relative permittivity and dielectric material. These parameters can vary the capacitance value of the variable capacitor.

- Change in dielectric constant can vary the capacitance of this transducer
- Area of the plates of these transducers can vary its capacitance value
- Distance between the plates can vary the transducers capacitance value

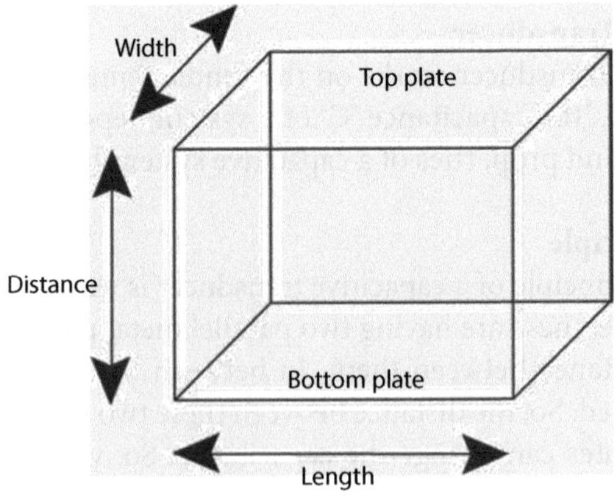


**Figure 5.50** Capacitive transducer.

In this method, the dielectric medium and area of the plates are kept constant. When the plates are moving then the distance is varied and this results in the changing of the capacitance of the capacitive transducer.

The above three methods are used to change the capacitance value of this transducer.

**Types of Capacitive Transducer:**
**The capacitive transducer are**

1. Parallel plate capacitance with rectangular plates
2. Cylindrical capacitor transducer
3. Semicircular parallel plates
4. Change in dielectric between parallel plates

**1). Parallel Plate Capacitance with Rectangular Plates**

This is also called a flat type of capacitive transducer. In this type of transducer, one plate is fixed and the other plate can be moved. By this variation, the distance or area can be varied. This results in the capacitance value of this transducer as shown in Figure 5.51.

If area 'A' varies and the capacitance value 'C' would be when the plates are having the distance of x, then

$$C = \varepsilon(A - wx)/d$$

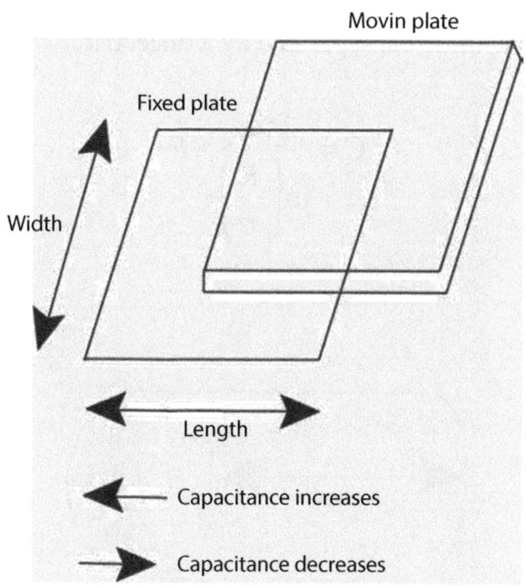


**Figure 5.51** Parallel plate capacitance.

If area 'A' varies and the capacitance value 'C' would be when the plates are having the distance of x, then

$$C = \varepsilon(A - wx)/d$$

If area 'A' varies and the capacitance value 'C' would be when the plates are having the distance of x, then

$$C = \varepsilon(A - wx)/d$$

If area 'A' varies and the capacitance value 'C' would be when the plates are having the distance of x, then

$$C = \frac{\varepsilon(A - WX)}{d}$$

**2). Cylindrical Capacitive Transducer**

A cylindrical capacitor is two concentric cylinders with a dielectric in between. Numerically substituting 'D' instead of R will give the same result. Mostly used to measure the angular displacement, the capacitance is maximum when the plates completely overlap, 180 degree is the maximum displacement of the capacitor. The cylindrical capacitor is shown in Figure 5.52.

The plates are cylindrical separated by a dielectric.

$$C = \frac{2\pi\varepsilon\ell}{\ln\left(\frac{R}{r}\right)} \times F$$

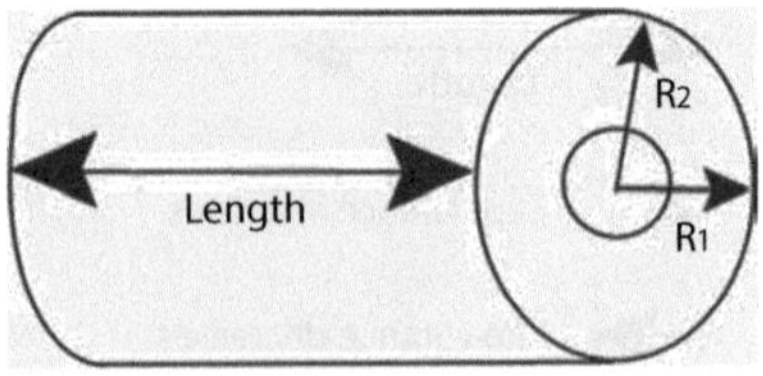


**Figure 5.52** Cylindrical capacitive transducer.

Where,

R is inner radius of outer cylinder
r is Outer radius of inner cylinder
l is length of the cylinder

**3). Semicircular Capacitive Transducer**

This type of capacitive transducer can provide the highest capacitance value when the two capacitive plates are overlapped to each other. These are preferable when the circuit requires the maximum capacitance. The semicircular capacitive transducer is shown in Figure 5.53.

In this type capacitive transducer, area $A=\frac{\pi r^2}{2}$ and the capacitance $C=\frac{\varepsilon \pi r^2}{2d}$.

**4). Change in Dielectric Medium between Parallel Plates**

Dielectrics are usually placed between the two plates of parallel plate capacitors. They can fully or partially occupy the region between the plates. When the dielectric is placed between the two plates of parallel plate capacitor, it is polarized by the electric field present. The surface charge densities are considered as $\sigma_p$ and $-\sigma_p$. When the dielectric is fully placed between the two plates of capacitor then it's dielectric constant increases from its vacuum value as shown in Figure 5.54.

When the dielectric medium varies between the two parallel plates of this transducer, capacitance of the transducer also varies.

The capacitance of a parallel plate capacitor is given by

$$C\frac{\varepsilon A}{d}\times F$$

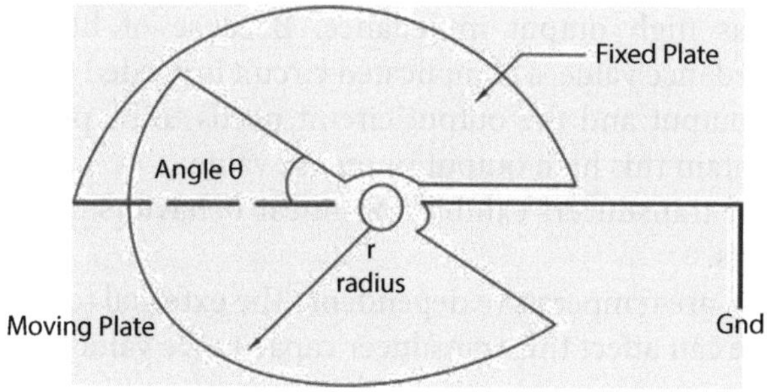


**Figure 5.53** Semicircular capacitive transducer.

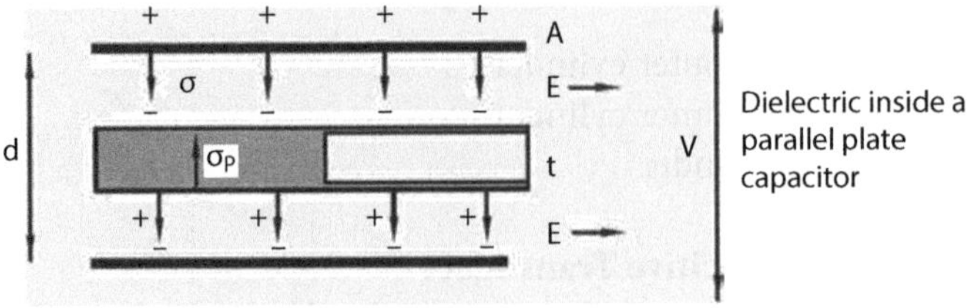


**Figure 5.54** Dielectric placed between two plates.

$$E = E_o E_r$$

Where,

$Eo = 8.85 \times 10^{-12}\ F/m$

$E_r$ is relative permittivity of material

A is cross sectional common area of plates

D is plate separation

The capacitance of the capacitor can be varied and change in its value can be used for transduction in a transducer.

The advantages of this method are:

1. These transducers offer high input impedance
2. The frequency response of these transducers is extremely high
3. These transducers are highly sensitive
4. These consume low power to operate and are therefore called low power consuming devices
5. High resolution can be possible by using these transducers

The disadvantages of this method are:

1. It has high output impedance. Because of high output impedance value, a complicated circuit is needed to measure the output and the output circuit needs to be powerful to maintain this high output immense value.
2. These transducers exhibit non-linear behaviors due to edge effects.
3. These are temperature dependent. The external temperature value can affect this transducer capacitance value.

The applications of this method are:

1. This transducer has a wide range of applications in determining the quantities like temperature, displacement, and pressure, etc.
2. These transducers have applications in the field of linear and angular displacement with the sensitivity factor.
3. One of the best applications of this transducer is to find the humidity level. As the humidity value changes the capacitance value of this transducer also changes. By this value, we can measure the change in humidity.
4. The variable capacitance pressure transducer is applicable to find the pressure variations by using the variable capacitance.

**iii. Inductive Transducer**

Inductive Transducer is the self-generating type, also called passive type transducer. The first type like self-generating uses the principle of fundamental electrical generator. The electric generator principle is when a motion among the conductor as well as magnetic field induces a voltage within the conductor [2, 23] The motion among the conductor and the field can be supplied by transforms in the measured. An inductive transducer (electromechanical) is an electrical device used to convert physical motion into modifying within inductance.

**Inductive Transducer Working Principle**

The working principle of an inductive transducer is the magnetic material's induction. Just like the electrical conductor's resistance, it depends on various factors. The magnetic material's induction can depend on different variables like the twists of the coil over the material, the magnetic material's size and the flux's permeability. The Inductive Transducer is shown in Figure 5.55.

The magnetic materials are used in the transducers in the path of flux. There is some air gap between them. The change in the circuit inductance can occur due to the air gap change. In most of these transducers, it is mainly used to work the instrument properly. The inductive transducer uses three working principles which include the following:

- Self-Inductance Change
- Mutual Inductance Change
- Eddy Current Production

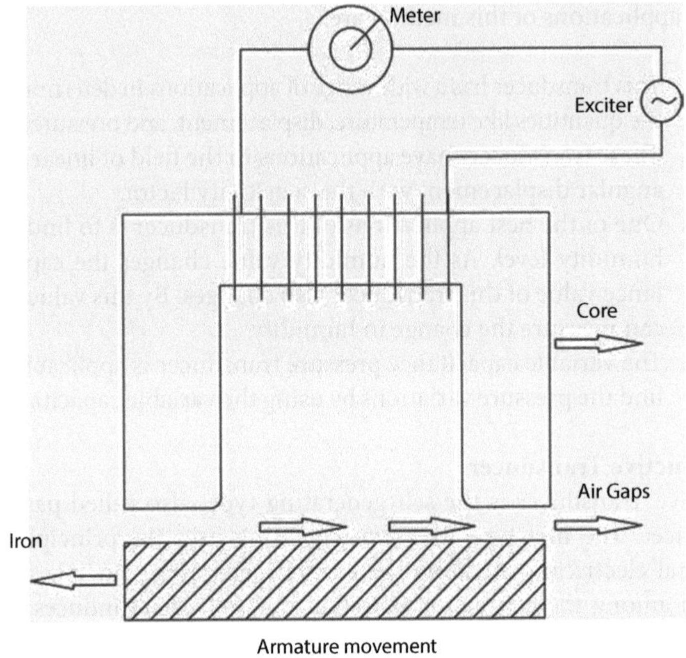


**Figure 5.55** Inductive transducer.

## Self-inductance Change

It is defined as the phenomenon in which a change in electric current in a circuit produces the induced electro motive force in the same circuit. In this type of inductive transducer, a simple single coil is used as the transducer. When the mechanical element whose displacement to be calculated is moved, then it will change the flux path's permeance which is generated from the circuit. It modifies the inductance of the circuit as well as the equivalent output. The circuit output can be directly adjusted against the input value. The self-Inductance image is shown in Figure 5.56.

The coil's self-inductance can be derived by

$$L \frac{N^2}{R}$$

Where,

N is the number of twists in the coil
R is the magnetic circuit's reluctance

**Figure 5.56** Self-inductance.

The reluctance 'R' can be derived from the following equation

$$R = \frac{1}{\mu A}$$

Thus, inductance equation becomes

$$L = \frac{N^2 \mu A}{l}$$

Where,

A is Coil's cross-sectional area
l is Coil's length
μ is Permeability
Geometric form factor $G = \frac{A}{l}$
then the inductance equation

$$L = N^2 \mu G$$

The self-inductance is changed by a change in the number of twists, geometric form factor 'G' and permeability 'μ'.

**Mutual Inductance Change**

In this type of transducer, two different coils are arranged. In the primary coil, the excitation can be generated with external power source whereas in the next coil the output can be attained. Both the mechanical input as well as output is proportional.

It work on the principle of change in mutual inductance. It uses several coils for the purpose of knowing. These coils include their self-inductance which is indicated by $L_1$ and $L_2$. The common inductance among these two twists can be derived by the following equation.

$$M = \sqrt{L_1} \cdot L_2$$

Therefore common inductance is altered by unstable self-inductance otherwise through the unstable coupling of coefficient 'K'. Here, the coupling coefficient mainly depends on the direction and distance among the two coils. As a result, the displacement can be measured by fixing one coil and making the secondary coil movable. This coil can move by the power source whose displacement is to be calculated. The change in mutual inductance can be caused by the change in displacement coefficient coupling distance. This mutual inductance change is adjusted by measurement and displacement.

**Eddy Current Production**

Whenever a conducting shield is located close to a coil carrying alternating current, then the current flow can be induced within the shield, which is known as "eddy current". This kind of principle is used in inductive transducers. When a conducting plate is arranged near to a coil carrying AC then eddy currents will be generated within the plate. The plate which carries eddy currents will generate their own magnetic field which works against plate magnetic field. So the magnetic flux will be reduced.

As a coil is located near to coil carrying AC, a flowing current can be induced within it, which in turn generates its own flux to decrease the flux of the current-carrying coil, and therefore a coil's inductance can be changed. Here, the coil is arranged nearer to the plate, and then high eddy current can be generated as well as a high drop within coil inductance. Thus, by changing the distance among the coil and plate, the inductance of the coil will change.

The applications of this method are as follows:

- This transducer is used in proximity sensors to measure position, touchpads, dynamic motion, etc.
- Mostly these transducers are used for detecting the kind of metal, to find lost parts, otherwise counts the objects.

- These transducers are also applicable for detecting the movement of the apparatus which include belt conveyor, bucket elevator, etc.

The advantages of this method are as follows:

- The responsively of this transducer is high
- Load effects will be reduced
- Strong against ecological quantities

The disadvantages of this method are as follows:

- The operating range will be reduced due to side effects
- The working temperature should be under the Curie temperature
- Sensitive to the magnetic field

**iv. Photoelectric Transducer**

The photoelectric transducer is used to convert the light energy into electrical energy. It can be designed with the semiconductor material [24]. When light falls on photosensitive element electric current is generated that is measured after amplification.

This transducer soak ups the light radiation which drops over the semiconductor material. The light absorption can boost the electrons in the material and therefore the electrons begin to move. The electron mobility can generate three effects like

- The material resistance will be changed
- The semiconductor's output current will be changed
- The semiconductor's output voltage will be changed

**Working Principle**

The working principle of a Photoelectric Transducer can be classified into photo emissive and photoconductive. In photo emissive type devices, once the radiation drops over a cathode it causes the emission of electrons from the cathode plane.

The diagram of the photoelectric transducer is shown in Figure 5.57.

The output of the PV cells can generate a voltage which is relative to the intensity of radiation. The occurrence of radiation can be IR (infrared), UV (ultraviolet), X-rays, gamma rays and visible light. In photo-conductive devices, the material's resistance can be changed once it is lit up.

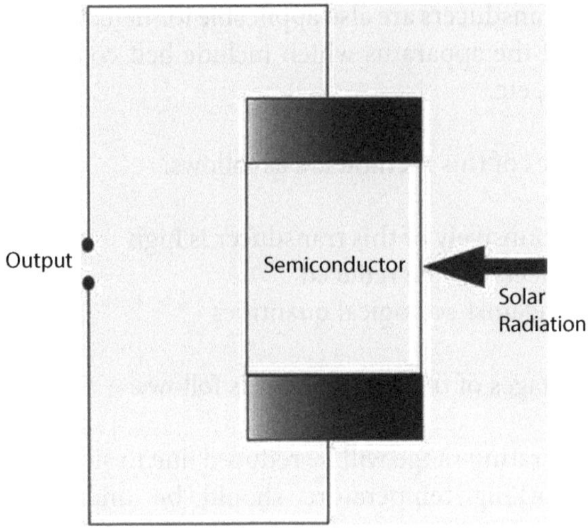


**Figure 5.57** Photoelectric transducer.

The applications of these transducers are as follows:

- These transducers are used in biomedical applications
- Pickups of pulse
- Pneumograph respiration
- Measure blood pulsatile volume changes
- Records Body movements

These transducers respond on electromagnetic radiation, dropping on the changing element's surface

These transducers are classified into five types:

1. Photo emissive Cell
2. Photodiode
3. Phototransistor
4. Photo-voltaic cell
5. Photoconductive cell

**Photo emissive Transducer**

When the photons of the incident light with sufficient energy incident on the surface, the electrons break their atomic bonds as well as the forces of the entire material lattices. This is called photo emissive effect [8].

1. Vaccum Phototube
2. Gas-filled Phototube
3. Photo multiplier tube

**1. Vaccum Photo Tube**

The surface specially shaped cathode is coated with some photo emissive material. The photo cathode is housed in a sealed glass tube along with another electrode called anode. The vacuum exists within the glass tube. The voltage is created between the photocathode and anode with the anode having positive voltage level while photocathode has a negative voltage level. The electrons are emitted from the photocathode and collected by the positive anode. If the intensity of light incident is stronger, a higher number of electrons are emitted. The Vaccum Photo tube is shown in Figure 5.58.

**2. Gas-Filled Phototube**

The tube enclosing photocathode and anode is filled with an inert gas such as argon. When light strikes the photocathode, the electrons are emitted and they are accurate towards the anode due to the voltage difference.

Similar in construction to a vaccum phototube except that it is filled with a noble gas (usually Ar) at a pressure of about 10 Pa. Photo electrons accelerated by the anode electrical potential ionize gas atoms. The additional electrons provide a substantial intrinsic gain. The Gas-Filled phototube is shown in Figure 5.59.

**3. Photo-Multiplier Tube**

It is the most widely used photo emissive device, having better characteristics to detect very low intensity light. It is an amplifying device. The beam

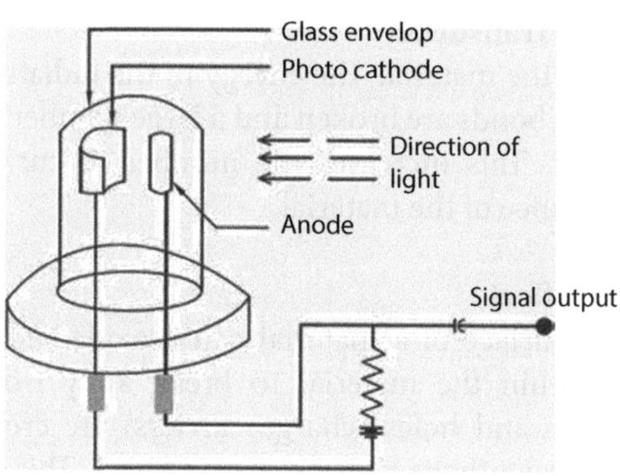


**Figure 5.58** Photoelectric transducer.

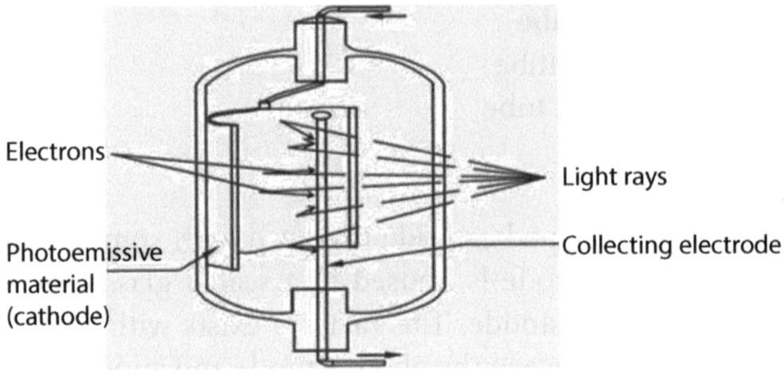


**Figure 5.59** Gas-filled phototube.

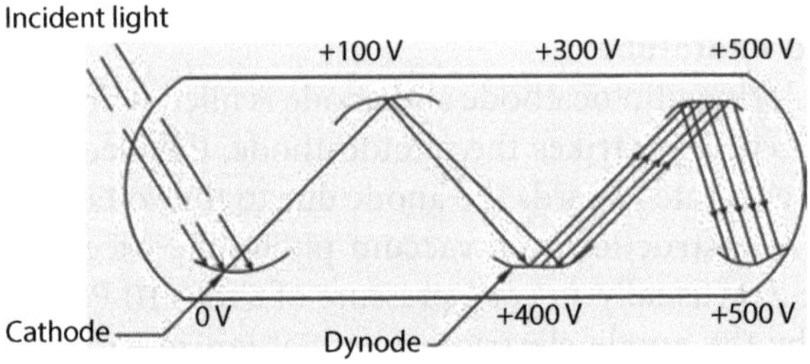


**Figure 5.60** Photo-multiplier tube.

of incident light is first made to strike a photo emissive material coated photocathode similar to the ordinary vacuum phototube. The electrons are emitted but the emitted electrons are not immediately drawn to the anode. The Photo-Multiplier Tube is shown in Figure 5.60.

### Photo-Conductive Transducer

When light falls on the material, the energy in the radiation increases the covalent bonds. The bonds are broken and a large number of electron-hole pairs are generated. This increases the number of current carrier and decreases the resistance of the material.

### Photoconductive Effect

Light striking the surface of a material can provide sufficient energy to cause electrons within the material to break away from their atoms. Thus, free electrons and holes (charge carriers) are created within the material and consequently its resistance is reduced. This is known as the Photoconductive effect.

**Working of Photo-Conductive Cell**

The Photoconductive cell is a two terminal semiconductor device. Its resistance will vary with the intensity of the incident light.

Light sensitive material is organized in the form of a long strip, zigzagged across a disc-shaped base with protective sides. The connecting terminals are fitted to the conducting material on each side of the strip; they are not at the ends of the strip. Thus, the light sensitive material is actually a short, wide strip between the two conductors. For added protection, a transparent plastic cover is usually included. The Photoconductive cell and symbol is shown in Figure 5.61 (a) & (b).

The Photo conductive cell use two material such as Cadmium Sulfide (CdS) and Cadmium selenide (CdSe). Both materials are change in light intensity. The light intensity and temperature sensitivity is important difference between the two materials.

**Light Intensity**

Cadmium selenium response time is around 10ms and the Cadmium Sulfide response time is 100ms. Both materials slowly changes in light intensity. The temperature sensitivity is another important difference between two materials. There is large change in the resistance of a cadmium selenide cell with changes in ambient temperature but the resistance of cadmium sulfide relatively stable.

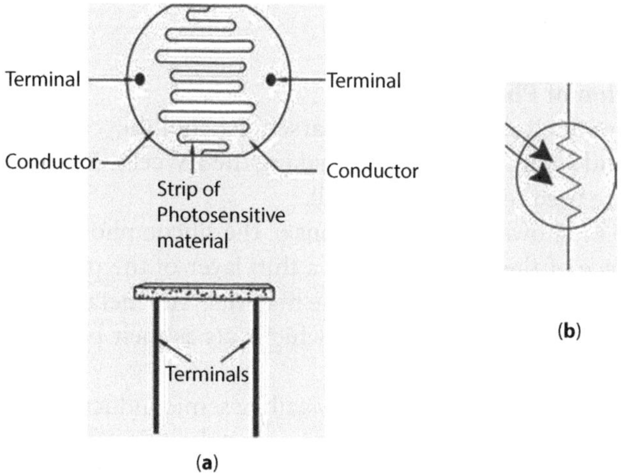


**Figure 5.61** (a) Photo conductive cell. (b) Symbol.

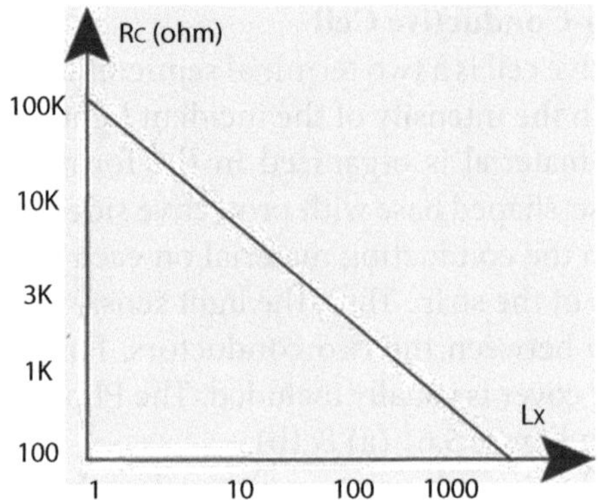


**Figure 5.62** Photoconductive cell Illumination characteristics.

**Characteristics of Photo Conductive Cell**

The illumination characteristics are shown in Figure 5.62. When the cell is not illuminated its resistance more than 100 kilo-ohms. This resistance is called the dark resistance. When the cell is illuminated its resistance less than few hundred ohms.

**Photovoltaic Transducer**

**Photo Voltaic effect**

When an open circuited P-N junction is illuminated a large number of electron-hole pairs are generated in the region near junction. A small voltage appears across its terminals, hence acts as a voltage source. This phenomenon in which light energy is converted to electrical energy is called photovoltaic effect.

**Construction of Photovoltaic Cell**

The semiconductor materials like arsenide, indium, cadmium, silicon, selenium and gallium are used for making the PV cells. Mostly silicon and selenium are used for making the cell.

Figure 5.63 shows the constructions of the silicon photovoltaic cell. The upper surface of the cell is made of a thin layer of the p-type material so that the light can easily enter into the material. The metal rings are placed around p-type and n-type material which acts as their positive and negative output terminals, respectively.

The multi-crystalline or mono-crystalline semiconductor material make the single unit of the PV cell. The mono-crystal cell is cut from the volume of the semiconductor material. The output voltage and current obtained from the single unit of the cell is much less. The magnitude of the output

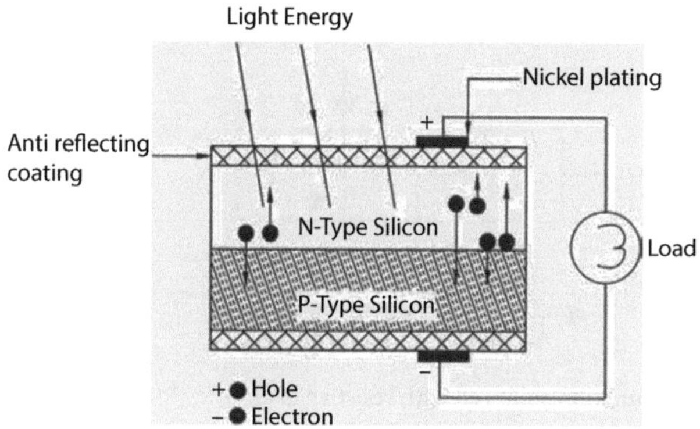


**Figure 5.63** Photovoltaic cell.

voltage is 0.6 V and that of the current is 0.8 V. The different combinations of cells are used for increasing the output efficiency.

**Working of PV cell**

The light incident on the semiconductor material may pass or be reflected through it. The PV cell is made of the semiconductor material which is neither a complete conductor nor an insulator. This property of semiconductor material makes it more efficient for converting the light energy into electric energy.

When the semiconductor material absorbs the light, the electrons of the material start emitting. This happens because the light consists of small energize particles called photons. When the electrons absorb the photons, they become energized and start moving into the material. Because of the effect of an electric field, the particles move only in the one direction and it develops the current. The semiconductor materials have the metallic electrodes through which the current goes out of it.

The Figure 5.64 shows the PV cell made of silicon and the resistive load is connected across it. The PV cell consists the P and N-type layer of semiconductor material. These layers are joined together to form the PN junction.

The junction is the interface between the p-type and n-type material. When the light falls on the junction the electron starts moving from one region to another.

Factors to be considered while selecting a transducer are as follows:

- It should have high input impedance and low output impedance, to avoid loading effect

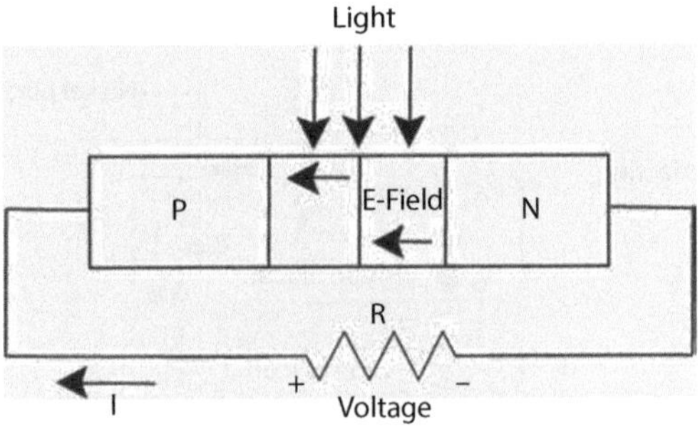


**Figure 5.64** P-N junction solar cell with resistive load.

- It should have good resolution over entire selected range
- It must be highly sensitive to desired signal and insensitive to unwanted signal
- It should be small in size
- It should be able to work in a corrosive environment
- It should be able to withstand pressure, shocks, vibrations, etc.
- It must have a high degree of accuracy and repeatability

The requirements of a good transducer are as follows,

- Smaller in size and weight
- High sensitivity
- Ability to withstand environmental conditions
- Low cost

**Piezoelectric transducer**

A piezoelectric transducer converts the physical quantity into an electric voltage. It convert mechanical stress into voltage. It is used for measuring the physical quantities like force, pressure, stress etc., The produced electric voltage can be measured using voltage measuring instruments to calculate the value of stress [25].

The piezoelectric material transducer uses the piezoelectric material which has special property known as the electro-resistive element.

**Types of Piezoelectric Materials**

Some of the types of piezoelectric materials are:

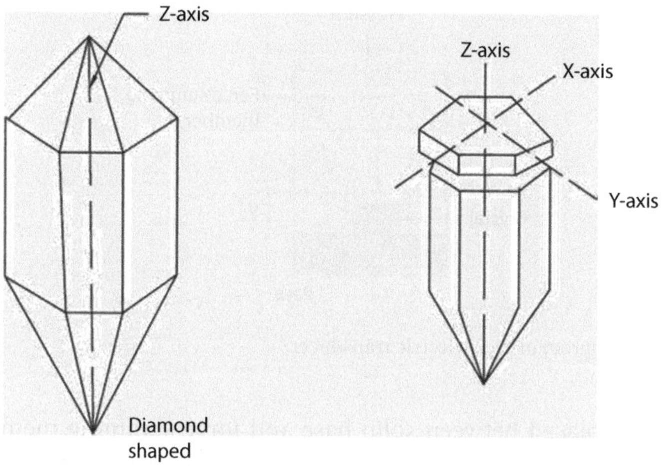


**Figure 5.65** Quartz crystal.

1. Natural Piezoelectric Materials are: Quartz, Rochelle salt, Topaz, Tourmaline-group minerals and some organic substances such as silk, wood, enamel, bone, hair, rubber, dentin.
2. Artificially manufactures piezoelectric materials are Polyvinylidene difluoride, PVDF or PVF2, Barium titanate, Lead titanate, Lead zirconate titanate (PZT), Potassium niobate, Lithium niobate, Lithium tantalate, and other lead-free piezoelectric ceramics. The Quartz crystal is shown in Figure 5.65.

The main substances exhibiting piezoelectric effect are quartz, Rochelle salts and Tourmaline. Rochelle salts show greatest piezoelectric activity, but mechanically they are weakest as they break easily.

**Basic Piezoelectric Pressure Transducer**

A piezoelectric transducer works with the principle of piezoelectricity. The face of piezoelectric material, usually quartz, is coated with a thin layer of conducting material such as silver. When stress has been applied, the ions in the material move towards one of the conducting surface while moving away from the other. This results in the generation of charge. This charge is used for calibration of stress. The polarity of the produced charge depends upon the direction of the applied stress. The polarity of the produced charge depends upon the direction of the applied stress. The diagram of piezoelectric transducer is shown in Figure 5.66.

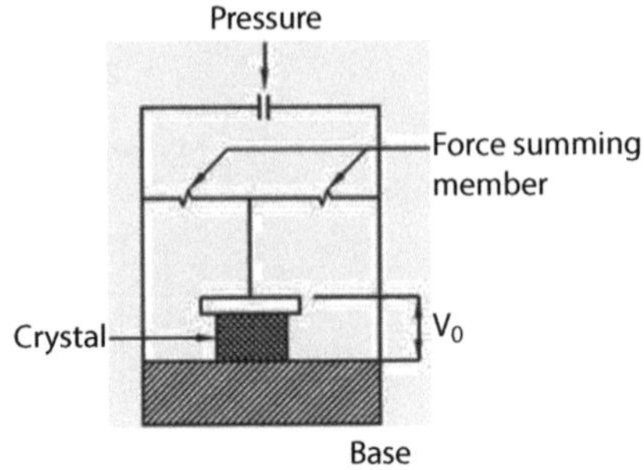


**Figure 5.66** Diagram of piezoelectric transducer.

A crystal placed between solid base and force summing member. The metal electrodes plated on to faces of piezoelectric crystal are taken out to the parallel plate capacitor.

$$V_o = \frac{Q}{C}$$

- The output is very high, about 1-30mV
- No external power is supplied is required
- High frequency response is excellent
- Its size is small and construction is simple

The practically the sensitive piezoelectric element of a piezoelectric pressure transducer is in the form of a pile of pairs of quartz discs. These discs are held in such a way that their optically flat faces are between a flat metal face called load plate.

$K_1$ is Stiffness of crystal
$K_2$ is Spring of Stiffness

$$P = P_1 + P_2 = K_1\Delta X + K_2\Delta X = (K_1 + K_2)\ \Delta X$$

$$\frac{P_1}{P} = \frac{K_1\Delta X}{(K_1+K_2)\Delta X} = \frac{K_1}{K_1+K_2}$$

$$= \frac{1}{1+\frac{K_2}{K_1}}$$

The applications of this method are as follows:

- Used for measuring surface roughness, in accelerometers and as a vibration pickup
- Used in seismographs to measure vibrations in rockets
- Used by automotive industries to measure detonations in engines
- Used in ultrasonic imaging in medical applications

The advantages of this methods are as follows:

- Rugged construction and small size
- High output with negligible phase shift
- Excellent frequency response

The limitations of this method are as follows:

- Temperature Sensitive
- Dynamic measurement only
- Water soluble

**Hall Effect Transducer**

In 1879 Physicist Edwin Hall discovered Hall-Effect Transducer. It measure a magnetic field and convert that measurement into voltage.

The transducer converts the magnetic field into an electric quantity which is easily measured by the analog and digital meters. The direct measurement of the magnetic field is not possible. Hence, the Hall-effect transducer is used [24].

The principle of the Hall-effect transducer is that the current carrying strip of the conductor is placed in a transverse magnetic field, then the EMF develops on the edge of the conductor. The magnitude of the voltage depends on the density of flux and this property of a conductor is called the Hall effect. The Hall-effect element is mainly used for magnetic measurement and for sensing the current. The metal and the semiconductor have the property of the Hall effect which depends on the densities and the mobility of the electrons. The Hall-effect transducer is shown in Figure 5.67.

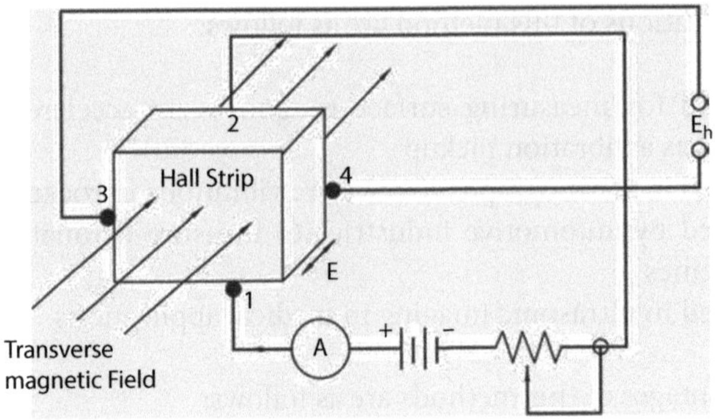


**Figure 5.67** Hall effect.

The current supply through the lead 1 and 2, the output is obtained from the strip 3 and 4. The lead 3 and 4 are at same potential when no field is applied across the strip. When the magnetic field is applied to the strip, the output voltage develops across the output leads 3 and 4. It develops the voltage which is directly proportional to the strength of the material.

The output voltage is,

$$E_H = \frac{K_H IB}{t} E_H$$

$K_H$ is Hall-effect Coefficient;

$$K_H = \frac{V-m}{A-Wbm^{-2}}$$

T is thickness of strip

Where,

I is current

B is Flux Density

The current and magnetic field strength can be measured with the help of the output voltages. The Hall-effect EMF is very small in conductors because of which it is difficult to measure. But semiconductors like germanium produces large EMF which is easily measured by the moving coil instrument.

**Applications of Hall-Effect Transducer**

**1. Magnetic to electric Transducer**
A semiconductor plate is inserted into the magnetic field to be measured. The magnetic line of force is perpendicular to the semiconducting plate. The transducer gives an output voltage which is proportional to B the magnetic field density.

**2. Measurement of Displacement**
It can be used for the measurement of the location or displacement of a structural element. The indirect acting position displacement or proximity transducer in case where a change of geometry of a magnetic structure causes a change of magnetic field strength. The measurement of displacement is shown in Figure 5.68.

**3. Measurement of Current**
The device serves to measure current in a conductor without the need of interrupting the circuit and without making electrical connection between the conductor circuit and meter. This is shown in Figure 5.69.

**4. Measurement of power**
Power measurements are made in high voltage circuits connecting the Wattmeter to the circuit through current and potential transformers.

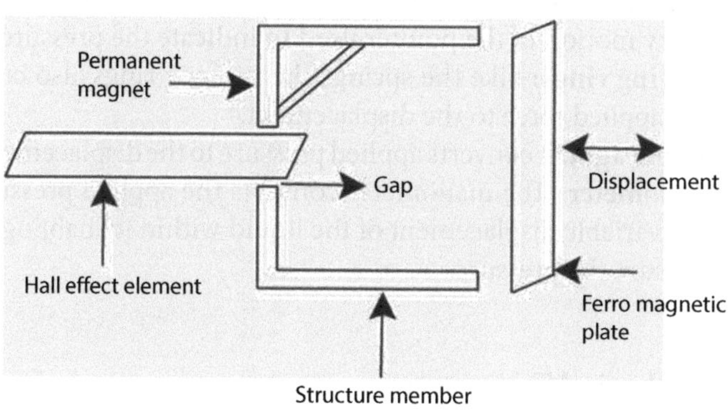


**Figure 5.68** Measurement of displacement.

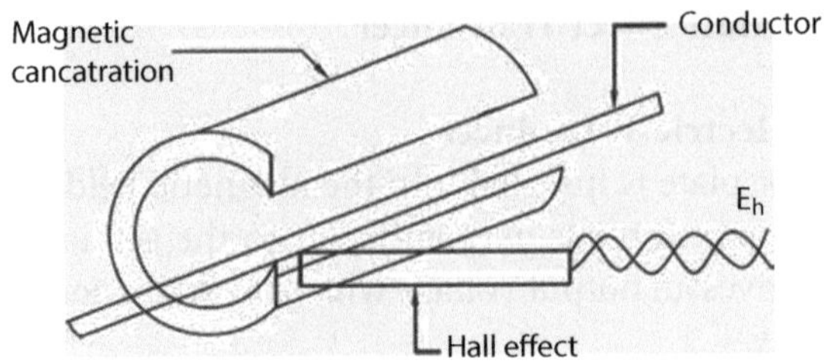


**Figure 5.69** Measurement of current.

**Mechanical Transducer**

If a transducer produces mechanical-nature signal as its output, then it is called mechanical transducer. Mechanical transducer are simple and rugged construction. They are accurate and operate without external power supplies but are not advantageous for many of the modern scientific experiments and process control instrumentation.

**Examples:**

Diaphragm, BourdonTube, Manometer, cantilever, gyroscope, thermocouple, Capillary tube etc.

There are number of mechanical transducers, some of the commonly used ones are described below:

1. **Bellows:** These are the elastic elements that convert the air pressure into displacement, and it is commonly used for the measurement of pressure.
2. **Bourdon tube:** This elastic tube converts air pressure to the rotary motion of the pointer used to indicate the pressure.
3. **Proving rings:** Like the springs the proving rings also convert applied force to the displacement.
4. **Diaphragm:** It converts applied pressure to the displacement.
5. **Manometer:** The manometer converts the applied pressure into variable displacement of the liquid within it enabling to measure the pressure.

**Types of Mechanical Transducer**

- Elastic Members (Diaphragms, Bellows)
- Thermal (e.g., Thermocouple, Bimetallic)
- Mass (e.g., Manometer, Seismic meter)

- Hydro-pneumatic
- Static (e.g., Float, Hydrometer)
- Dynamic (e.g., orifice, venture, pitot tube)

## Student and Lecturer Conversation

**1. Lecturer: What is meant by measurement?**
**Student:** Measurement is an act or the result of comparison between the quantity and a predefined standard.

**2. Student: What are the basic requirements of measurement?**
**Lecturer:** The basic requirements of measurement are:

i. The standard used for comparison purpose must be accurately defined and should be commonly accepted.
ii. The apparatus used and the method adopted must be provable.

**3. Student: What are the functions of a measurement system?**
**Lecturer:** A measurement system consists of a transducing element which converts the quantity to be measured in an analogous form. The analogous signal is then processed by some intermediate means and is then fed to the end device which presents the results of the measurement.

**4. Lecturer: What are the functional elements of the measurement systems?**
**Student:** The three main function elements of the measurement systems are:

I. Primary Sensing Element
II. Variable Conversion Element
III. Data Presentation Element

**5. Lecturer: What are the different types of Systematic Errors?**
**Student:** The errors are divided into three categories:

I. Instrumental Errors
II. Environmental Errors
III. Observational Errors

**6. Lecturer: What is an Instrument?**
**Student**: It determines the value or magnitude of a variable.

**7. Lecturer: What are the types of instruments?**
**Student:** The types of instruments are:

I. Mechanical Instruments
II. Electrical Instruments
III. Electronic Instruments

**8. Lecturer: Why must Instruments is calibrated?**
**Student:** Calibration of all instruments is important since it affords the opportunity to check the instrument against a known standard and subsequently to find error and accuracy.

**9. Lecturer: What is a Standard?**
**Student:** A standard is a physical representation of a unit of measurement. The term standard is applied to a piece of equipment having a known measure of physical quantity.

**10. Lecturer: What are the different types of standards?**
**Student:** The standards are:

I. International Standard
II. Primary Standard
III. Secondary Standard
IV. Working Standard

**11. Lecturer: What are Electrical Instruments used for?**
**Student:** Electrical instruments are used to measure electrical quantities like current, voltage, electric power, electrical energy, power factor, frequency, etc.

**12. Lecturer: What is a Multi-meter?**
**Student:** A Multi-meter or a Multi-Tester, also known as a VOM (Volt-Ohm-Milli ammeter) is an electronic measuring instrument that combines several measurement functions in one unit.

**13. Lecturer: What is a Cathode Ray Oscilloscope (CRO)?**
**Student:** A CRO is an electronic device with a CRT as its main component and other associate circuits consisting of a power supply unit, a saw tooth-wave generator, horizontal and vertical amplifiers.

**14. Lecturer: For what purpose is a triggering circuit provided in a CRO?**
**Student:** In a CRO, a triggering circuit is provided for synchronizing two types of deflections so that horizontal deflection starts at the same point of the input vertical signal each time it sweeps.

**15. Lecturer: Which part is called heart of CRO?**
**Student:** Cathode Ray Tube

**16. Lecturer: What is the function of the probe in CRO?**
**Student:** The probe performs the very important function of connecting the test circuit to oscilloscope without altering, loading or otherwise disturbing the test circuit.

**17. Lecturer: What is sweep time?**
**Student:** It is a time period during which the beam is swept from left to right on the screen of a CRT by the linearly increasing saw tooth waveform.

**18. Lecturer: What is a digital oscilloscope?**
**Student:** The design of a digital oscilloscope is centered around an analog to a digital converter. Once the signal is digitized, it is processed by an FPGA or a microprocessor before being displayed.

**19. Lecturer: What is Threshold?**
**Student:** When the input to a transducer is increased from zero, there is a minimum value below which no output can be detected.

**20. Lecturer: What is a Static characteristic?**
**Student:** The characteristics of quantities or parameters measuring instruments that do not vary with respect to time are called static characteristics. The parameters may vary slowly with respect to time:

I. Accuracy
II. Precision

III. Sensitivity
IV. Resoltuion
V. Threshold
VI. Linearity

**21. Lecturer: What is the accuracy of a measurement?**
**Student:** The degree of closeness of a measurement compared to the desired value.

**22. Lecturer: What is precision of a measurement?**
**Student:** A measure of the consistency or repeatability of measurement, i.e., successive reading does not differ.

## References


1. Robert B. Northrop, *Introduction to Instrumentation and Measurements,* 3rd Edition, CRC Press, 2014.
2. Shawhney A.K., *A course in Electrical and Electronics Measurement and Instrumentation*, Dhanpat Rai & Co., 2015.
3. Alan S. Morris and Reza Langari, *Measurement and Instrumentation: Theory and Application,* 2nd Edition, Academic Press, 2015.
4. Lindley, David, *Degrees Kelvin: A Tale of Genius, Invention, and Tragedy,* pp. 132–133, Joseph Henry Press, 2004 ISBN 0309167825.
5. Nineteenth-century Scientific instruments. London Berkely: Sotheby Publications University of California Press, 1983.
6. A. V. Bakshi, U. A. Bakshi, *Electrical and Electronic Measuring Instruments,* First Edition 2009.
7. Edward Hughes (revised by Ian McKenzie Smith). *"Electrical & ElectronicsTechnology"* Pearson Education Limited. Indian Reprint 2002.
8. S.K. Singh, *Industrial Instrumentation and control,* second edition, Tata McGraw-Hill Education, 2007.
9. B. C. Nakra, K. K. Chaudhry, *Instrumentation, Measurement And Analysis,* second edition, Tata McGraw-Hill Education, 2006.
10. Kuldeep Sahay, Shivendra Pathak, *Basic Concepts of Electrical Engineering,* New Age International Publishers, New Delhi, 2006.
11. Abhijit Chakrabarthi and Sudipta Debnath, *Basic Electrical Engineering,* McGraw-Hill Education, New Delhi, 2018.
12. Tony R. Kuphaldt, *Lessons In Electric Circuits, Volume I - DC,* First Edition, Koros Press, London, 2012.
13. A. D. Helbincs and W. D. Cooper, *Modern Electronic Instrumentation and Measurement Techniques*, PHI, Singapore, 2007.
14. Ian Hickman, *Digital Storage Oscillocope*, Newnes, London, 1997.

15. Peter H. Sydenham, *Handbook of Measurement Science,* John Wiley & Sons, Chichester, 1983.
16. Francis T. Farago and Mark A. Curtis, Handbook of Dimensional Measurement, Industrial Press, Fourth Edison, New York, 2011.
17. John G. Webster and Halit Eren, *Measurement, Instrumentation, and Sensors Handbook,* CRC Press, Boca Raton, 2014.
18. Alan S. Morris, *Measurement and Instrumentation Principles,* Elsevier, United Kingdom, 2001.
19. D.V.S. Murthy, *Transducers and Instrumentation,* PHI, NewDelhi, 2004.
20. Walt Boyes, *Instrumentation Reference Book,* Fourth Edition, Butterworth-Heinemann, Oxford, 2009.
21. Luigi Fortuna, Salvatore Graziani, Alessandro Rizzo and Maria G. Xibilia, *Soft Sensors for Monitoring and Control of Industrial Process,* Springer, Singapore, 2007.
22. S. Vijayachitra, *Transducers Engineering,* PHI, New Delhi, 2016.
23. M. A. Laughton and D.F. Warne, *Electrical Engineer's Reference Book,* Newnes, London, 2002.
24. G. R. Jones, Electrical Engineer's Reference Book, Fifteenth Edison, Newnes, London, 1993.

# Index

Printed in the USA/Agawam, MA
January 3, 2021

767483.031